Turn Your Idea OR Invention INTO Millions

BY Don Kracke

ALLWORTH PRESS
NEW YORK

08   07   06   05   04   7   6   5   4   3   2

Published by Allworth Press
An imprint of Allworth Communications
10 East 23rd Street, New York, NY 10010

Book design by Jelly Associates, NYC

Page composition/typography by
SR Desktop Services, Ridge, NY

Library of Congress Cataloging-in-Publication Data
Kracke, Don.
Turn your idea or invention into millions / by Don Kracke.
        p.   cm.
ISBN 1-58115-019-9
1. New products—Management.   2. New products—Marketing.
    3. Inventions—Economic aspects.   I. Title.
HF5415.153.K73   2001
658.5'75—dc21
2001004551

Printed in Canada

TO **Lisa, David, Ted,** AND **Tina.**

Table OF Contents

v

Foreword

"Most people will admit that they have envisioned at least one good innovative product or service. . . . But usually, hopes of turning this idea into a reality are quickly dashed when faced with the labyrinthine process of bringing a product to market. . . . Overwhelmed by the lack of answers, many give up before they really start."

This is a quote from my "Speedmarketing™" report written over a decade ago. Here we are today, and while progress has been made, the existing void of knowledge and lack of networking channels in the field of innovation continue to be the greatest blocks to an individual's ability to get his or her ideas into the marketplace.

That's why I created (invented) the Invention Convention tradeshow approximately a decade and a half ago. And why, as Executive Director of The National Congress of Inventor Organization (NCIO), it's my driving purpose to help nurture the spirit of creativity by providing the much needed information, networking, and educational resources.

That brings us to the story about how this book became "born-again." A few months back when I was speaking at the Mark Taper Auditorium on "The Business of Inventing," hosted by the L.A. Central Public Library—Science, Technology, and Patents Department, I had a sudden realization.

I had invited Don Kracke, my friend, to be my guest speaker that day. And while we were up on the stage answering questions, it hit me that there I was with the author of *How To Turn Your Idea Into A Million Dollars,* one of the best books ever written about the process of getting your new ideas to market, i.e., researching, patenting, manufacturing, funding, and marketing options for launching new products.

Even though I've known Don for some time now, it was at that split second that I actually grasped how truly instrumental his book had been in helping people with their ideas. This book has been long out of print, but it was a best seller in the late seventies, and I've known that many of the underlying principles and information in it are timeless.

At the time he wrote it, there was very little practical information on how to market a new product. In fact, I had often recommended Don's book to budding inventors and entrepreneurs in my own consulting practice.

He probably doesn't even realize how much his book has influenced the field of inventing. His book was literally a gold mine of information with case histories about real people who had been there, done that, and who had first-hand experience in getting their ideas to market.

For several years I had encouraged Don to republish this classic. Again, it was at that moment that I became convinced that this book had to be reprinted. I told him that I would be happy to act as his literary agent to see if we could find a publisher. He said to go ahead, so I made some calls, did the pitch, and the result is this book in your hands.

I told Don I would be happy to help edit the updated book. He said, "While you're at it, why don't you give me a foreword?" Naturally, I was more than honored.

I'm sure you'll agree with me that, Who better to guide you than Don Kracke, a former ad agency executive, who began his phenomenal inventing career with the Rickie Tickie Stickie? This stick-on decal, which was designed to look like a daisy during the Hippie/flower-child era, became the defining symbol of that generation (along with Volkswagen's bug and bus). In fact, over 90 million of those bright and colorful flower decals were sold in just one year.

Kracke parlayed his great success with the stick-on daisy decals into an internationally acclaimed new product design and development company based in Palos Verdes, California. Here are just some of his accomplishments:

* His various companies have created and marketed over 2,300 new products over the past three decades.

* Over $1 billion in retail sales have been generated from new product ideas he has had a hand in creating.

* He has negotiated and signed over 700 licensing agreements.

* Don has also consulted numerous international companies, including The Coca Cola Company, Campbell Soup Company, General Mills, Rubbermaid, Corning, and Mattel, to name just a few.

* He is an award-winning artist with 217 national awards who calls himself a "designasaur."

* Don has appeared on numerous television shows, including *The Oprah Winfrey Show, CBS Evening News,* and *Nightline.*

Which brings me to the discovery of Don's secret to his phenomenal success. Back in 1998, when he was a Keynote Speaker at the Invention Convention, the topic of failure was that year's theme—due to the presence of another speaker, Robert McMath, known as "Mr. Failure" (because of McMath's collection of over 60,000 failed products, which he houses in a museum).

Kracke revealed that turning failure inside out was his magic formula. Don says that statistically he fails 97 percent of the time in getting his brilliant, new product ideas to market. But he utilizes failure as a learning tool, a springboard to success, never losing faith in his ability to produce and keep swinging in anticipation of the next home run.

He should know. In 1990 alone, he received almost $2 million in royalty income.

Yet for the 2,300 products he has created and which have gotten to market, he says there were somewhere around 97,000 that didn't make it!

Don inspires hope. He says, "As long as you maintain your sense of enthusiasm, regardless of how often you get shot down, your chances [of succeeding] are every bit as good as mine."

Anybody who has had success in any field has chalked up numerous failures in comparison to their few successes. Risk-taking and learning are part of the success process, because for every failure, you're that much closer to success.

From Don's experience of this, he's been know to say, "I've been crapped on so many times, I ended up with a fertile mind."

So one thing's for sure—"going for it" means that you'll have to take some B.S. along the way. But like Don, remember that it's the tenacity, perseverance, stick-to-itiveness, and learning from your failures that will lead you to the height of success.

And you won't be alone; you'll have Mr. Kracke himself guiding you in this inspiring, enlightening, and entertaining book, *Turn Your Idea or Invention into Millions.*

And now, as you start your journey, I leave you with an appropriate quote on commitment by Johann Wolfgang von Goethe (1749–1832):

> Until one is committed, there is hesitancy, the chance to draw back, always ineffectiveness. Concerning all acts of initiative (and creation), there is one elementary truth the ignorance of which kills countless ideas and splendid plans: that the moment one definitely commits oneself, then Providence moves on.
>
> All sorts of things occur to help one that would never otherwise have occurred. A whole stream of events issue from the decision, raising in one's favor all manner of unforeseen incidents, meetings, and material assistance, which no man could have dreamed would have come his way.
>
> Whatever you can do or dream you can, begin it. Boldness has genius, power and magic in it. Begin it now.

With Godspeed,
Stephen Paul Gnass
Executive Director, National Congress of Inventor Organizations, and Founder, Invention Convention (www.inventionconvention.com).

1

Modesty BE Damned

*You are about to read the
finest book ever written on
the subject of getting
your brilliant new product
ideas to market.*

WELCOME. AND CONGRATULATIONS. I MEANT WHAT I SAID. YOU ARE about to read the finest book ever written on the subject of getting your brilliant new product ideas to market. Yes, I said the finest. It was the finest in its original format some years ago. It's even better today. But, we'll get to more on that subject when we reach the chapter called "One-Man Band Must Toot Own Horn."

Right now, I want to set the record as straight as possible; this is the definitive book on licensing your wondrous new product ideas to a manufacturer on an advance-against-royalty basis. Any alternative to licensing is a very, very distant second. I want you to clearly understand that I am going to show you how to go into *The Idea Business*, not the manufacturing business. If all else fails, we'll discuss your options where manufacturing and selling your idea on your own is your only option. But that will be a long time from now.

The second most important point to glean from this epic is the fact that I absolutely and positively want you to forget about your million-dollar ideas. The title for this book was my publisher's idea. It's possible to turn your invention or idea into a million dollars. In fact, I've created several million-dollar ideas in the past few years. However, I'd really prefer that you have a bunch of $200,000 ideas working at any given moment. What you're going to experience by the time we're finished is the wonderful sensation that you're a bottomless pit of creativity. And, most importantly, you'll know how to harness that creativity and head it in the directions that will give you the best chance of generating a bit of income from your genius. Believe me, it's a bona fide rush when you walk down the aisle of a store and see one of your creations sitting on a shelf waiting to be purchased.

Note my reference to "one of your creations." I am positive that if you have one idea, one patent, or one invention, you will have many, many more right behind that first one. And that really is the essence of what we're going to explore together. As I said, forget that once-in-a-lifetime, million-dollar idea. Let's get you started on the path to creativity, intelligently connected to the reality of the marketplace. That is what I do as well as anyone.

Incidentally, throughout this book, I'll be using real, hard-dollar numbers when discussing my successes and failures. I hate with a passion the parade of pretenders that hang around the invention/idea business and profess to be authorities on getting ideas or inventions to market even though they can't tell their SKU from their shoe.

And finally, let's put to rest that one big question that has been buggin' you since you read the cover notes which referred to the original version of this book. It was called *How to Turn Your Idea Into a Million Dollars*. The first words out of Regis Philbin one morning when I was doing his show were, "Don, are you a millionaire?" Now, I was in front of a zillion viewers, the IRS, my ex-wife, and God only knows how many creditors. Fortunately my brain beat my mouth to the punch. "I wrote the book on how to make a million dollars, not how to keep the million dollars," I replied.

Actually, for once in my life I don't owe anyone a lot of money. My wife and I are comfortably semi-retired. And, if push really came to shove, I guess my net worth would just squeak up over the one-million-dollar mark.

This book and my whole career as an inventor was catapulted as a result of my experience with the success of the Rickie Tickie Stickies craze of the 1960s, which I'm going to share with you in the next chapter.

Don Kracke
Lake Arrowhead, California

IT'S Rickie Tickie™ Time

② $$$

You'll find that the recurring theme in this book is, "Don't do what I did. Do as I say."

YOU'RE GOING TO LOVE THIS CHAPTER. IT CONTAINS ALL THE REALLY good stuff that a creator of any new product dreams about when asleep or awake. We're talkin' over a million dollars in sales in the first year. We're talkin' becoming mini-famous enough to do *Oprah, Nightline,* and *Regis.* We're talkin' about Rickie Tickie Stickies. And we're talkin' reruns. With their attendant residuals, thirty years later.

FIFTEEN MINUTES of FAME—PART I

Since most of you reading this weren't born when I had my first fifteen minutes of fame, let me take you for a brief trip-trop down memory lane. To like *even before Austin Powers.* Politically, in the United States, times were tense. Because of the Vietnam war, the opportunity for a movement of sorts to start—the hippies—was extant. The antiwar "peace-niks" used flowers as their symbol of hope. Flowers were in. Doves were out. Rainbows were soon to be on their way.

Since most of the hippies were poorer than church mice, those that could afford any kind of vehicle drove your basic $350 Volkswagen Beetle. Then they were called Bugs. Now, they're called Slugbugs. Go figure. And since all those VWs looked alike, the more enterprising hippies, also known as flower children, would hand-paint flowers and peace symbols on their bugs. Great concept. Rotten artwork.

One day, my then-wife and I were driving on an L.A. freeway when we passed three hand-decorated VWs in a row. "There has to be a better way," I mused. From there we discussed stencils and spray paint but discounted that approach as too permanent and too messy. How about pressure-sensitive decals? Clean, neat, easy to use. Nonpermanent. We decided to go for it. That was a good idea unto itself.

The amazing fact is that over ninety million flowers were sold in the first year alone. Of those, my company sold 11.7 million. The rest were sold by the knockoffs. For those of you who don't know that term, it aptly describes a whole gaggle of companies that copy other people's ideas. We'll talk later about ways to foil those villains. For the here and now, however, suffice it to say that the major lesson I learned from having manufactured and sold Rickie Tickie Stickies on my own was: An idea is one thing, distribution is everything.

Before I go any further, here are the financial facts I promised you. Of my $1,300,000 in first-year gross wholesale sales of Rickie Tickie Stickies, I netted (on paper) $400,000 in before-tax profits. Neither my wife nor I were able to take any salary for the first year. And in those days, the Feds took fifty percent of profits for taxes. But since all our profits were in receivables and inventory, I had to borrow $200,000 to stay out of jail. I've often thought jail might have been the more pleasant choice.

But rather than focus on things dour, let's use my great good luck of the Rickie Tickie Stickies adventure to set out a few positive results, like my being able to go on a leave of absence from my day job and my not having gone back after thirty years.

Another positive is the fact that I never, ever doubt myself when what seems like a really dumb idea pops into my head. "Who knows?" I mutter to myself. "This could be another Rickie Tickie Stickie."

And then there's the very real satisfaction one feels when one has acted on an idea instead of just sitting around complaining that someone else has probably created your idea in the first place. So why bother? This very point is the reason I stopped accepting invitations to speak to Inventors Groups. All those bozos wanted to do was to figure out how they'd react were they to be hypothetically knocked off. Talk about misdirected effort.

Another positive that evolved from the Rickie Tickie Time was that folks who had originally discounted my enthusiasm for the idea business started to stop by, seeking advice.

And probably the greatest benefit was in my gaining the respect of my four kids. Success has a wonderfully positive effect on one's ultra-jaded teeny boppers. Which leads to a very important point regarding your involving whatever family group surrounds you. Involve them. Lay out all your hopes and dreams. Share your plans. Seek their advice. Include them because their help and understanding will be needed many times along the way to the successful launch of your Rickie Tickie–type idea.

And, most importantly, *do not ever fall in love with your ideas or yourself.* For your pocketbook's sake, be a helluva lot more critical than I when evaluating each of your new genius-level concepts. Do that prior to investing anything more than your time and energy. As an example, directly behind my desk where I'm writing this book are four three-inch-thick, three-ring binders. Each book contains notes and drawings covering about forty new-product ideas waiting patiently to be taken to market. The good news is that I have that many. The bad news is that I've spent exactly $324,209.63 in hard-dollar, out-of-pocket expenses developing that lump of creativity. And now that I'm seventy years old, the prospect of my screwing up my enthusiasm one more time and going out there and selling all that wonderful new stuff is almost more than I care to contemplate. Given a choice, I'd rather have my money back. So, if you're taking notes, mark up one for your being extremely critical of each and every new idea that comes-a-visiting your noggin.

FIFTEEN MINUTES of DUMB

I've been asked many, many times, since I am such an outspoken advocate of licensing one's ideas instead of manufacturing them, Why didn't I license my Rickie Tickie Stickies to a decal company or a gift company? There's only one honest answer: I did not know enough at the time to have successfully pulled off a licensing deal. Hell, I couldn't even spell licensing. In fact, at one point early on in my Stickie adventure, I was presented with an outstanding opportunity to license my entire operation to a company called Chicago Printed String (CPS). Dumb name. Great company. Jack Kuhlberg, the then-president of CPS, had heard from his sales manager and his gift-sales reps that our little California decal company's sales were going through the roof in the Los Angeles area. He called and asked if he could come visit. "Of course. Be our guest," was our friendly reply.

5

By the day Jack arrived at our office, we had sold a little over 40,000 packages of flower decals at $1.00 per package wholesale. Not a lot, but a remarkable number to someone as knowledgeable as Jack. Those 40,000 packages had been sold in a little over three months in the worst gift-selling period of the year. We were on a roll and didn't even know it. Jack's offer was very neat and clean; he offered to pay me for all our inventory at our cost. That was about $15,000 at that moment. He also agreed to pay me for my $25,000 in current receivables. And to make the deal irresistible, he offered to pay me a $10,000 signing bonus as well as retaining me to design my new product-line extensions for the soon-to-be international Rickie Tickie, Inc. And, oh yes, he had started by offering me a 5 percent royalty.

Had I had this book to read, I would have snapped at the offer. It was fair, and I was pooped beyond my ability to continue my two-full-time-jobs life. The problem was that at the time I had no idea Jack's deal was fair. So I fiddled away over a month trying to figure out just how I was being screwed. I managed to take so long that, in the meantime, Jack had sold his company and the new owners couldn't care less about our little, pip-squeak operation. And as a final aside on the subject, had I sold to Jack, his sales would have been in the $8 million at-wholesale range because, as I mentioned earlier, *an idea is one thing, distribution is everything.* And Jack had the distribution worldwide.

The other important business truth I developed out of that period of my life was, *You can fund failure, and you can fund mediocrity, but watch out for great success.* You will never know what stone silence really sounds like until you ask your banker for a million-dollar line of credit when your personal net worth is $42,628 and your new company's assets come out a negative number.

The good news is that the Rickie Tickie adventure was very valuable in that it was the first time I realized how much better off I would have been had I made a licensing deal with CPS. Just for openers, 5 percent of eight million comes out a neat $400,000 in royalties. And, importantly, I would have had the time to shelter a bit of that income in an investment that would have helped generate income to me in my semi-retirement of today. I did ultimately sell Rickie Tickie Stickies to a Minneapolis company for a deal that would have made me rich in the following two years had that forty-four year old company stayed the course. Two years after my signing on, its ship sank. All the way to the bottom of chapter 7. Happily, out of more than 700 licensing deals I have made since then, only three others have gone belly up.

I DID MY HOMEWORK

And, before I forget, prior to my attempting to write this book, I carefully reviewed the existing books that generally cover the topics I'm including here. There are ten different books included in our appendix for your reference. Each has some valuable information regarding the invention trade. None, in my opinion, hits the mark as regards the real-world attitudes one must adopt in order to succeed in the idea business. Until you've been at the craft long enough to ripen your sense of humor, you probably would never tell an audience, "Philosophically, I'm positioned somewhere between Alfred E. Newman and Pinocchio."

And, until you have been knocked off by thirty-four different manufacturing and marketing companies on one new product idea, you'll have no true understanding as to how really helpless the average inventor is in such shameful situations. To that point, one of my major complaints with all ten of the books I reviewed is that each spends most of its time focusing on the technical minutia of the idea business and practically no time on the very vital creative aspects. The reason is simple; it's easy to focus on objective details. It's harder than hell to isolate subjective creative processes. And it is here that I have a tremendous advantage over almost everyone else writing on this subject. I have learned every important concept I will be sharing with you from actual experience in the invention trenches. As of this moment, I am the proud holder of twelve patents, fourteen registered trademarks, and 1,206 copyrights. Add to that the 708 licensing agreements I've personally negotiated and signed, and you end up with a large lump of real-life experience. And, believe me, real-life examples are always the best platform upon which to build any learning experience.

The major point that should stick in your gourd is the fact that I want you to focus on being in the idea business, not the manufacturing business. You could not have a better example than my mishandling of the opportunity to sell Rickie Tickie Stickies.

And lastly, you'll find that the recurring theme in this book is, "Don't do what I did. Do as I say."

Sailing Jacket FOR Idiots

Everything you will ever have to know about getting your invention or idea to market.

THE BEAUTY OF THE LITTLE STORY I'M GOING TO RELATE TO YOU NOW is that it incorporates every aspect of getting one's wonderful new product from The Idea Stage to The Cashing of the Royalties Stage. And, most importantly, everything in between is detailed as well. I can vouch for the authenticity of the facts because I lived each and every minute of the entire program. In person. And, unfortunately, unrehearsed.

Included here are all of the steps to be followed in getting one's great, new product idea to market. Let's start with a simple outline of those steps:

1. Having the idea
2. Reducing that idea to practice
3. Doing your homework
4. Doing a patent search
5. Protecting your idea
6. Preparing your prototype and other attendant presentation materials

7. Photographing your idea professionally

8. Developing your hit list of potential manufacturer licensees for your wonderful new-product idea

9. Calling to set up your in-person presentation

10. Making presentations in person

11. Maintaining your dignity while being crapped on by experts

12. Biting the bullet and never giving up

13. Negotiating your contract

14. Keeping track of your royalties

15. Watching your idea exceed the average 4.2 year Kracke product life span by well over 5 years

16. Having the rights to your idea revert back to you for future use

17. Seriously considering relicensing your idea to some small, start-up Action Sportswear company

As you can readily see, this chapter could be a book in itself. So, I'll do my best to keep from rambling. Strike that. I'm incapable of not rambling. However, I will give this my best shot.

DRINKS all AROUND

Our story starts back in my serious drinking days. Which explains why I was agreeable to crewing on the Newport Beach, California to Ensenada, Mexico annual sailboat race for nine years in a row. Actually, I hated the sailing part, but I loved getting to Ensenada each year. Ensenada is the home of one of the craziest bars on this planet. It's called Hussong's Cantina. I remember walking for nine years in a row into Hussong's for a wee taste of a tequila shooter. To this day, I never once remember leaving. However, that's really not the point of this tantalizing and exciting story. The point is that each and every year I'd have to try and remember which side of the boat was port and which was starboard. I couldn't even remember the little poem that was written especially to help old jerks like me remember port and starboard.

Chugging home after about the sixth race, it struck me that a sailing jacket that was half red and half green would be a big help in solving my left/right dilemma—especially if I were to stencil the words "port" and "starboard" in giant letters on each applicable sleeve. With my stroke-of-genius idea, as long as I faced the pointy end of the boat, I'd be able to tell port from starboard.

Immediately, I made a little note detailing my idea and dating it. Being a professional drinker in those days, I needed all the memory-jogging help available. And, sure enough, it was almost a week after getting home that I wore the same pants as I had on the sailing trip. They had been washed with my port/starboard note still in one of the hip pockets. The poor note was faded but readable.

Remember that little note part so that when I finally get you up to having twenty or thirty ideas working simultaneously, you'll be sure to remember all the new ones that will keep cropping up. I'll make you a bet here and now—there will be no stopping those ideas from buggin' you.

Please go back to the outline now and check off the *having the idea* step. Believe me, that was the very easiest part.

What we're about to do next is *reduce the idea to practice*. There are all kinds of opinions as to what the phrase "reducing to practice" really means. In my book (which this is) it means starting a notebook with nonremovable pages that includes a description of the idea and any attendant drawings I might have developed. It especially means dating and having some nonrelative witness that dating. It's that simple. The hard part, for me at least, is keeping my trap shut and not blabbing to everyone about how clever I am and how great my idea is. More quiet at this point is more better.

So now it's on to the *doing your homework* part. At this point in my port/starboard saga, I created the little non sequitur that may help you in the future: "Happiness for the inventor is looking for what isn't there and then not finding it." That's, in essence, what doing your homework is all about. You literally go from store to store looking to see if you can find any product being sold at retail that is anywhere near what your idea entails. A word of caution here: Don't cheat and go to Pep Boys Auto Supply stores if your idea involves ladies lingerie. We all know you don't want to find your great, new idea already on the market. However, if you do find something anywhere near your concept already being sold, you will save a ton of time and money by giving up and moving on to another idea. Only if you were forced into the manufacturing and distribution process would "working around" an existing product make any sense at all. And even then, that approach would be closer to nonsense than you'd ever care to be.

If you're lucky and don't find anything like your idea on the market, be sure to get the names of manufacturers who are making a similar product. Also get the city, state, and zip code. All that information is on the product's label.

In order to keep my recounting of the port/starboard jacket story on track and moving right along, I'm going to condense my six weeks of shopping to the simple statement of fact that I could not find anything in any retail sailing-gear-type stores that even slightly resembled my idea. That was a happy conclusion that logically led to the next step in the process, which was the *doing the patent search* part.

THE CONCURRENT DEVELOPMENT SYNDROME

Since, in those days, I didn't have a clue as to what "doing the patent search" meant, I'll presume that you don't have a clue either. This is a good presumption because this is the single most important step in the entire new product development process. My having had literally hundreds of patent searches turn up ideas nearly identical to mine has led to my creating the master cliché of all time: *Just because I thought of it does not mean it's original.* As you can already tell by the tone of my writing, I'm an egoist in an ego-driven business. And the bigger my ego grows, the more original and clever I presume I am. Any idea that pops into my gourd must be original. Right? Wrong. *Just because I thought of it does not mean it's original.*

Right about here you're probably wondering why I put so much importance on the originality factor. As a reminder, the primary goal for my initial effort in each and every new product adventure is to license my idea to a manufacturer on an advance-against-royalty basis. Unless I am dealing in an original concept, I have nothing to bring to the table. And no manufacturer is going to pay me a big advance against a giant royalty just because I'm pretty.

There are only two options available to you if you agree with me that a patent search is a necessity in any inventor's life. You may do the search yourself or you may have a patent attorney do it for you. The attorney will charge you anywhere between $300 and $900, with the average being about $500. You can do a patent search yourself for free. However, saving money at this point in the game can be very expensive because you do need a legal opinion from a patent attorney before filing a patent application.

An example of one effort to save money that backfired was the time that my son David (a brilliant attorney) decided to do his own patent search to cut corners. Since a majority of his schooling and a portion of his everyday job involved research, he was comfortable with the work to be done. As an aside, most major cities in the United States have a U.S. Patent and

Trademark Depository Library (PTDL) where copies of all patents are on file. David lived in Portland. He located the Oregon PTDL in the nearby Salem library.

At that time, all patents issued were on microfiche and anyone could look at them all day long for nothing. They're filed, logically, by classifications. David did not find any patents anything like his idea. Armed with that information, I licensed his idea to a company in Illinois. As part of that company's due diligence, their patent attorney had their own search done. David's idea had been patented previously by someone else. Even though our intent was honorable, this new information clouded the issue to a point that our deal fell apart. David simply had not gone back enough years to find that applicable patent during his search. That explains why I always spend the money and have my patent searches done professionally.

Incidentally, Stephen Gnass has pointed out to me that as of 2000, the Patent and Trademark Office has put all of the patents ever issued in the United States since 1790 on its Web site at *www.uspto.gov*. Talk about information overload. Now anyone can do a preliminary patent search without leaving home. If one has a computer, that is.

Also, Stephen added, there are roughly eighty PTDLs in the United States with staff members who are happy to guide you in doing a patent search on your own. However, both Stephen and I think that making this your only and final patent search is a dumb idea. I think that when an inventor tries to save money by doing stuff himself, that generally, such false economies lead to screwups. The patent search, as I've said a hundred times, is the most important part of the entire creative process if, in fact, you're planning to license your idea. So, please, don't leave it to an amateur like yourself. And, in passing, were you to retain a patent attorney to get you a utility patent and he does not recommended a patent search, get another attorney.

At this point, I must interrupt the flow to address a major point of difference I have with my longtime friend and patent attorney, Scott Kelley. Scott is very good at his job. He's also very honest, which prompts him to advise me not to pay for a patent search when applying for a design patent. His logic is impeccable. The patent office will search the category as part of its due diligence regardless of our having done the same thing. So why not save the $800 or $900? The only reason is timing. Even a simple design patent process generally takes eighteen to twenty-four months to complete, so I would have to wait that long before trying to license my idea to a manufacturer. That's too long to wait for my kind of volatile new-product ideas. Or for my temperament. And, since I have nothing to present until I know my idea is clearly original, having that information in a timely manner (read six to eight weeks) is essential to my method of presenting and licensing ideas.

I'd better get back to my outline, or this chapter will become a book, too. Our patent search did not uncover any similar ideas already in the process on the port/starboard jacket, so I moved forward with the project. I did not apply for a design patent at that time. I had opened a docket with my patent attorney, and I had what I needed, which was the information that my idea was apparently original. I always put off moving further along the patent process until I have some interest from a manufacturer. That's because I currently have 164 new-product ideas in development. Of those, I only paid to search the most promising. And, as a matter of course, when I get a manufacturer interested in licensing my idea, part of the advance negotiations involves who will pay to complete the patent procedures.

I JUST LOVE PROTOTYPES

Armed with the extremely vital information that my port/starboard jacket idea was indeed original, I forged ahead to the phase I always enjoy the most. That's the *preparation of the prototype* part. For those of you who don't know the word, a prototype is nothing more than a three-dimensional model of what you hope will become a finished product. Please note my point about three-dimensional models. Whenever I prepare presentation materials, I work in the third dimension rather than in two-dimensional drawings. It's more expensive, but you'll be amazed how much you learn about your new idea once you see your baby and are able to hold it. Or wear it.

The prototype for the port/starboard jacket cost me $20 for a neighbor to make the shell jacket. I did the stencil work myself, so I was in dirt cheap on this idea.

So now I have a prototype. Now what? Now we pay real money to get professional photographs of our prototype. Each of the hundreds of prototype photos I have had shot cost about $200 each. And each photo has been worth the investment time and time again.

Incidentally, I just reread my first draft of this chapter, and I discovered that I have been entirely too cavalier regarding my references to money. From this point forward, I'll assume you're flat broke. Regardless, from this point forward, we will be tightfisted in our approaches to getting your ideas to market. Which reminds me, we were talking about my port/starboard jacket and how valuable good photos are to your presentations. For example, whenever I am doing a new product presentation, I always include suggested packaging ideas (in 3D color) that incorporate my target licensee's packaging format. And yes, I change it each time I resubmit to the next company. A bit silly. But effective. Good photos obviously help in the preparation of these important packaging and display items.

When I referred to "other attendant presentation materials" in item six in this chapter's outline, I was mainly thinking of the new-product names I try to develop to support every one of my presentations. Throughout this book you'll find references to Rickie Tickie™ Stickies, Walter™, Neat Garage™, Dandy Cans™, Fancy Cans™, Bagzilla™, Laurelwood™, and many, many others. We'll discuss in depth what those little TMs mean in our legalese chapter. What they mean in the context of your preparing for your presentation is that you are bringing more to the party than just your new-product idea. As Mister Confucius often said, "The more you bring to the party, the larger the advance." Simple.

HAVE a HIT LIST

We're currently up to the *developing of your hit list of potential manufacturer licensees* for your now highly developed, properly protected, and cleverly designed new-product idea. You'll recall I collected manufacturer's names during my homework stage. I went back to the stores with sporting jacket departments and started checking my notes of the manufacturer's name, city, and zip code. In those days, White Stag was the dominant manufacturer in the category. Catalina and Empress also appeared throughout my search. As another aside, while you're in the stores, ask the department managers which manufacturers (in your category) they like to work with. The best use of this tidy bit of knowledge leads us right into the next phase of our program, which is *calling to set up your in-person presentation.*

This, for most of you, will be the most difficult phase to work through. It's the most difficult because a phone call is, in fact, a form of public speaking. And, as you probably know, besides death, public speaking is the most feared activity in which a human being ever gets involved. However, if you use, almost verbatim, the dialogue I used when first calling White Stag, you'll be light years ahead in the appointment-setting game. This, of course, follows my having called Portland information (remember that label note) and asking for the phone number of the corporate offices of the White Stag Sportswear Company.

"Good morning, The White Stag Corporation, how may I direct your call?" the very, very cheery receptionist chirped. (Remember, those were the good old days when real people actually answered the phone. Today, once you get through all the push buttons and talk to a real, live operator, the technique is the same.)

"Good morning. My name is Don Kracke. That's spelled K-r-a-c-k-e. (Do the spelling part slowly. It makes you sound more important. And be sure to use your own name.) I'm a professional inventor calling from Los Angeles. Please give me the name of the person at White Stag who is most responsible for reviewing new-product ideas from outside inventors."

At this point there invariably is a long pause followed by something like, "I'm sorry, I'm new. I don't know if we have such a person working here."

"That's no problem. Would you please give me the name of your vice president of marketing and/or sales?"

Another pause. "That would be our Bill Snyder. He's at extension 222."

"Thanks a lot. And, by the way, what's Mr. Snyder's secretary's name?" Pause. "Trixie? Is that with an *ie* or a *y*? Y. Thanks again. And, what's your name? Carmen? Thanks, Carmen. Goodbye." (Note: At this point you hang up. Do not ask to be transferred. You will now call White Stag again and follow the mechanical voice's instructions which will include your punching in extension 222.)

"Hello, Trixy with a *Y*. This is Don Kracke with an *E*. I'm a professional inventor calling from Los Angeles. Sally Smith at The Haughtery Department Store gave me your guys' name. I would like to talk to Bill to set up an appointment to show him a new product I've developed that would fit perfectly into the White Stag line. Incidentally, I have done a patent search, and there's nothing like my idea on the market. I also have a patent pending, so you'll be in the clear as far as reviewing the idea goes." Trixy connected me to Bill, and, in essence, I repeated the same opening salvo to him that I had used on Trixy. (Coincidentally, Bill was also responsible for the development of new product at White Stag, so I was talkin' to the right guy.) Besides being interested, Bill was also very cordial. In fact, twenty years later he's still cordial, and we're still friends.

At this point, I must emphasize the fact that the purpose of your calling is to *set up an in-person interview*. Do not, repeat, *do not* settle for anything less. At least in the early stages of this critical part of your odyssey.

Your goal (as was mine with Bill) is to set the in-person meeting at a major airport closest to your hometown. You will do this (as I did with Bill) by asking what would be the next time he would be passing through, let's say for example, Los Angeles? Bill said he'd be in San Francisco the next week, so we settled on meeting at United's Red Carpet Club, of which Bill was a member in good standing. One reason for my not wanting to fly to Portland was to save the airfare. The second reason is really far more important. You always want your presentation to be made outside of your target's office if that is at all possible.

I walked into the Red Carpet Room wearing my prototype jacket which caused a guy across the room to start waving at me. I knew I was about to meet the famous Bill Snyder. Cordial Bill startled the hell out of me by demanding that I take off "that damn jacket." That was before we had even shaken hands. "Why?" I asked in all innocence. "Because one of my competitors might be in the room, Jerko," Bill replied. I did as he had demanded. Bill's next comment rocked me even more. "How'd you know?" he asked. "How'd I know what?" I countered. "How'd you know White Stag was going into the sailing jacket segment of the business next year?" We settled on serendipity and got on with our get-acquainted and presentation meeting.

Bill took the jacket with him after promising not to show it to anyone outside of White Stag. (His paranoia was contagious.) He also asked me if I would like to design the packaging for my new port/starboard line to be introduced by White Stag the following year. I was on cloud nine and climbing towards ten when he agreed to a $2,500 advance against a 5 percent (of wholesale) royalty. "I'll check with our president, and we'll get an agreement and a check off to you in a couple weeks," were Bill's parting lines. I was very close to incontinent.

I used those two weeks to further ingratiate myself to Bill by creating full-color, three-dimensional packaging and catalog layouts. He called me personally when my layouts arrived by mail in Portland. "Trixy and I love the designs," was all I remember his saying. Cloud eleven drifted a hundred feet beneath my wings.

CURSES! SCREWED AGAIN!

Then an unfunny thing happened to me on my way to the bank. Neither the agreement nor the check arrived. Bill did not return my phone calls as promptly as he had at first. Even I could tell something was happening. And it was not good. Try three months of nongood. Bill, who was still as sold on the jacket idea as the day we met, finally had to tell me his president had decided that my idea did not fit into the marketing plans of the White Stag

Company at that time. Years later, Bill admitted to me that my idea was just too off the wall for his stodgy, old-line corporation's mindset.

Was I disappointed that my lead-pipe-cinch deal fell apart? I was furious. For about ten seconds. At least the idea had not been turned down because it smelled like a skunk. "Does not fit into the marketing plans of our company at this time," is a phrase I've heard at least a hundred times over the years. "Your idea stinks," I have never heard.

So, encouraged by my belief that White Stag was wrong and I was right, I looked up the phone number of the Catalina Swimsuit Company. They had a small Action Sportswear Division, so they were at least a reasonable target. The real plus was that they were local. In L.A.

The same phone routine as with White Stag was followed with Catalina. Again, I ended up with the vice president of marketing who also doubled as the director of new-product development. I was unsuccessful in my attempt to get him to visit me at my office. (Remember the distraction theory.) So I trip-tropped over to his place. I knew I was at the right spot when the VP's secretary laughed out loud upon my entering the executive office suite of Catalina wearing my prototype. "I've got to get one for my Dad," she said. "He's a sailor."

The presentation was very comfortable and ended on the high note of the vice president's requesting that I leave the jacket prototype and the "beautiful packaging ideas" for a few days. You're right. The night before, I had changed the White Stag logo to the Catalina logo on all the layouts. Another custom presentation, by Kracke!

My request for a $2,500 advance against a 5 percent (of wholesale) royalty was "very realistic" according to my dear, dear friend, the VP. "We'll be back to you within two weeks," was his parting remark. I was already into this project six months, so what was another couple of weeks' wait? Especially when you're waiting for $2,500.

Three days later my entire presentation arrived back at my office via UPS. The note simply said that, "Upon further review, we decided your port/starboard jacket concept does not fit into our marketing plans at this time."

NOBODY SAID it WOULD be EASY

Was I discouraged? You bet your sweet bippy I was. (Note the clever use of an expletive generic to that era.) More to the point, I was pissed. Regardless, the next name up was the Empress Floatation Sailing Jacket Company. And Halleluya, they, too, were in L.A. I used the same phone routine and ended up speaking to the president of Empress. You'll find out that with the smaller companies, it's not unusual for the president and/or the spouse to be in charge of the New Product Development Division. I went over the next day wearing my prototype jacket as usual. Unfortunately, there was no receptionist to chuckle at my cleverness.

However, the president of Empress heard me and came out to greet me. He took one look at my jacket, went "Harumph" (or something close to it), and, as he was returning to his office muttered, "I don't think that's the least bit funny! The floatation-jacket business is a very serious business." Not as serious as I was driving home, I'll tell you. But, at least I knew where I stood with the folks at Empress, and by the end of the day I was looking up the number for The Mighty Mac Sailing Jacket Company in Massachusetts. I called the next morning and the kindly voice that answered the phone said, "That is I," when I got to the "Who is the most responsible person for reviewing new-product ideas from outside inventors?" part. "I" turned out to be the president/receptionist and director of New Product Development. He had owned Mighty Mac for many years and, of course, he'd be delighted to review my new-product idea as long as he was assured that I was protected. Now that was a twist. A caring client. I assured him I was protected and added the fact that he, too, was protected by my protection. It sounded like a condom commercial.

Off went my great idea. The Empress logo had been changed to the Mighty Mac logo. I did not fly to Boston to make an in-person presentation because, frankly, I could not afford it. As it turned out, my being there would not have helped one little bit. The moment my parcel arrived at Mighty Mac, the President called and was very complimentary as regards my presentation. "Yours is one of the most professional new-product presentations I've ever seen," he stated with great enthusiasm. "By the way, I did my port/starboard jacket ten years ago." I couldn't believe my ears. I asked him what he meant. He said he meant what he had said. "I did my port/starboard jacket about ten years ago. What's not to understand?" To this day I don't remember what I muttered to Mr. Mighty Mac during the next few moments. I did manage to get the point across that I'd like to see a picture of *his* port/starboard jacket. And sooner would be better. "Sure," he said. "And I'll send your stuff back, too. My port/starboard jacket didn't sell worth a damn so I'm not the least bit interested in ever trying to sell one again."

I was greatly relieved when I received a photo of the Mighty Mac version. The body of the jacket was all white. Above each pocket were the words Port and Starboard neatly embroidered. And, as is often the case, Mr. Mac did not bother to patent his idea since he was producing it himself. That explains why nothing showed up in my search. However, the experience was traumatic enough that I set the whole project aside and went back to my day job.

ALL BREAKS are NOT BAD

That day job took me to a meeting at Sears Roebuck in Chicago the following summer. At a break, several of us were looking at the sailboats on Lake Michigan. The joys of sailing became the topic of the day. Which gave me an opening to tell my port/starboard jacket story in almost as much detail as I just related it to you. "And what did it look like?" a lady from Sears Marketing asked. I dug into my briefcase and produced the photo I had taken a year and a half earlier. Why I had that photo amidst all the real business reports is still beyond my understanding. Again, serendipity. Regardless, everyone laughed at my cleverness, and the marketing lady asked if I had ever shown the idea to the Land's End Catalog Company. Not being a real sailor but rather a lover of tequila, I had to admit I'd never heard of Land's End. "Best sailing gear catalog in the world," someone chimed in. "The very best," another Sunday sailor agreed. "And their offices are over on Ogden Avenue," a third added. "Why don't you call and see if they're interested," a fourth musketeer volunteered. "Why the hell not?" I added with a bit too much bravado. "Why the hell not?"

My phone call to Land's End followed almost verbatim the ones I used on White Stag, Catalina, Empress, and Mighty Mac. As luck would have it, Gary Comer, the founder and president of Land's End, was having his lunch at his desk that day and, yes, he'd be happy to hear about my great new sailing jacket idea as long as it was protected with a patent pending. And why didn't I come right on over? I was there in fifteen minutes.

Gary took one look at the photo of me wearing the prototype and started chuckling. "Can I use this photo in my catalog?" "Of, of, of course," I stammered. "What kind of a deal do you want?" Gary asked. "I'd like a royalty on each jacket sold," I said tentatively. "What kind of royalty?" Gary asked. "I don't have a clue," I countered and added, "I've never licensed a garment design in my life." *When all else fails, the truth works wonders.* I continued by asking Gary what he normally paid in royalty on his sailing jacket line. "I don't have a clue," was his honest answer. "I've never made nor sold a jacket in my life." We're talkin' about the blind leading the really blind here. "Why not add a dollar a jacket to whatever selling price you decide on?" I asked hopefully. "Sounds fair," said Gary. And we shook hands on the deal. So much for complex negotiations.

A few months later, my picture was popping off the pages of about eleventy-zillion Land's End catalogs. And better yet, in a little over three months after the catalog arrived at the Land's End customers' homes, I received a check from Gary for $1,298, which covered the 1,298 jackets he had sold in that first quarter. Each year for nine years, Land's End sent me anywhere from $4,800 to $6,500 in royalties. The total royalty income came to a little over $47,500. Because of perseverance and patience, coupled with a bit of dumb luck, I was well on my way to becoming a professional inventor. Over time, I realized that were I to have about twenty Land's End–type deals working at any given time, that extra $100,000 per year would certainly help keep the wolves from the door.

GET ME the IRS!

Regarding *keeping track of your royalties*, do what I say, not what I did. Which was deposit each quarter's royalty check and forget it except for tax purposes. That laissez-faire attitude led to my missing an entire year's royalties from Land's End until their new comptroller was told by my bookkeeper that we had a verbal agreement. They had no intent of cheating me out of my royalties. There was just no record anywhere that they owed me any money. Three days after the discovery, I had a check for $4,937, which covered the previous year's sales. So, for heavens sake, keep a little notebook or something and record your royalty payments. A letter of agreement wouldn't hurt any either.

My reference to the nine-year run of my port/starboard jacket is only an academic trivia point; I've tracked the sales life of each of the over 2,300 different consumer products I've created. *The average life of the no-brainer, decorative, novelty-type stuff I do is normally about 4.2 years. My port/starboard jacket ran over twice as long as that average.* Which, in turn, led me to points 17 and 18 of the outline.

I have taken back the rights to my port/starboard patent (yes, it did issue) from Land's End since they were no longer interested in selling them. And, since ten years have passed since the jackets were on the market, it might be time to try again.

Regardless of what happens next, the experience was the inspiration for the lead sentences in my original version of this book. "Overnight success takes about a year. Or longer." And that's as true today as it was then.

HOW TO Have AN Idea

Ideas don't just happen. You have to work for them. And there are some tricks you should know.

I'LL MAKE YOU A BET RIGHT HERE AND NOW. I'LL BET YOU THAT BY the time you finish reading this book, you'll have an idea you won't be able to resist taking to market. I've seen it happen over and over again. I'll be babbling along to someone trying to explain to him how to have ideas and suddenly he'll get this funny look on his face. I'll become aware that he is no longer listening. He's just had an idea. So will you.

I'll give you an example, which may be telling tales out of school since it involves a book editor I know. I was talking to her about how many people have ideas and how little most of them know about what to do next, what steps to take after that. She said she was fascinated by the whole process and wished she could come up with a great invention, make a million dollars, and retire. She wondered how people go about having all those great flashes of genius.

"Nothing to it," I said. "You want to have an idea right now? A good one?"

She smiled. "Sure, why not?"

"Okay, how many manuscripts do you get in here every week? Dozens? A hundred?"

"A lot," she said. "I probably look at a thousand manuscripts a year."

"And how do they come to you?" I pressed. "I'll bet most of them are all wrapped up in brown paper with string and tape. You can probably tell that someone with about eleven thumbs wrapped it up on the kitchen table, right?" (Writers never seem to be very good at wrapping things. Or tidy, either. I understand that Thomas Wolfe delivered the manuscript of *Look Homeward, Angel* to his editor, Max Perkins, in a big corrugated cardboard grocery box. There were thousands of sheets of messy paper, words scribbled in the margins, blots and splotches all over everything.)

Sally, my editor pal, allowed that it was true, that the manuscripts did seem to have that worried-over-and-bundled-up-carefully-in-too-much-of-everything look.

"Uh-huh. And when you want to send the manuscript back to the writer, what do you do?" I asked rhetorically. "You probably send it down to the mailroom where somebody else who has about eleven thumbs hauls out the brown paper and string and tape and wraps it up again."

"Something like that," Sally admitted.

"Okay, how about having some kind of foldable cardboard box, just the right size for a book manuscript. You buy it flat, pop it open, stick the manuscript in it, close it with some locking tabs and mail it off. Push-pull, click-click. Just like that. The manuscript stays nice and neat, the package looks tidy and professional, and it only takes about thirty seconds for you to pack up the whole thing and be done with it."

"Hey," she said, "that might work!"

Of course it would work. It would probably even sell. Maybe not a million dollars' worth, but enough to bring in a nice royalty. By the time you read this, somebody may already have the mailer on the market, Sally, for all I know. In fact, I was just informed by my editor here at Allworth Press that manuscript boxes are now on the market. So there you have it.

THREE EASY LESSONS

Anyway, there are three lessons to be learned from the example of the manuscript mailer.

First, there's more to inventing something than meets the eye. If you think about it carefully—and Sally did—you'll realize that she didn't really invent anything at all in our conversation!

"Hey, wait a minute," she said. "I didn't invent that mailer. You did!"

"Nope," I said. "Neither of us did." You see, simply figuring out that something needs to be invented—a manuscript mailer, for instance—doesn't constitute inventing it. Just because you have a notion that a foldable mailer would be a good idea, don't think you've actually invented it. You're a long way from that, yet. You have to work out all the details of how it would function, how it would be manufactured, how much it would cost, where it would be sold, how it would be distributed, and how many you would expect to sell—to name just a few of the chores you still have to do before you have invented anything.

There are a lot of people wandering around this earth firmly convinced they have invented something, when all they've really done is figured out some new product that *needs* to be invented. That's daydreaming, not inventing.

The second lesson to be learned from our example of Sally's manuscript mailer is that you don't have to be a specialist to come up with a valid, marketable idea. All it takes is a little bit of creative thinking—the kind you use everyday for other purposes. There's no mystery to it. You can do it. Your kids can do it. Your dimwitted neighbor can do it.

But probably the most important lesson to be learned from our example is the third one: You don't have ideas by sitting around and waiting for the muse of inventions to land on your shoulder. If you're going to wait around for a fit of inspiration to strike, you'd better bring your lunch. It's going to be a long, long wait. You've actually got to set out to invent something, lay out the ground rules and the information and then try to work out the invention. Getting started shouldn't be too hard. You can just look around you and you'll see a million things that need to be invented. Think about the problems you run into everyday, then solve them.

For instance, you probably have a paper towel dispenser in your kitchen. And you probably hate it as much as I do. When I try to tear off a paper towel, I generally wind up with a towel and a half. Or two towels. Or the whole roll falls into the cat's water dish. And it looks tacky, a little piece of cheap plastic amid all the shiny chrome and enamel of the kitchen appliances. There has to be a better way.

Or how about your steam iron? Every time you try to fill it with water, you dribble it all over the counter and it spills down the side of the iron. Then little hidden droplets of water collect in the nooks and crannies and when you start to iron with it, the water droplets spill out and make a big, damp blot on the thing you're ironing. When it's time to empty the iron, you make more steam than the Orient Express pulling out of Istanbul. And, of course, there's always a little water that won't come out. It stays inside the iron, rusting out the delicate innards so that every now and then a belch of brown gunk comes out on your ironing along with the steam. There has to be a better way.

I could give you more examples, but I think you get the idea. There are millions of things standing around waiting to be invented. All you have to do is sit down, figure out what they are, and then invent them.

GET to KNOW some BUYERS

There are people who make their entire living either coming up with ideas or developing the ideas of other people. They work for companies that live or die on the basis of their new products. Toy companies, for instance, are almost totally dependent on a new crop of great ideas every year. Buyers for major chain stores are constantly involved in the development of new items to add to their line of wares. A friend of mine is one of those latter people. You'll meet him from time to time in this book because of that fact and because he's articulate and intelligent about his work. As a housewares buyer for a major national chain, he's developed some tricks of the trade that could be valuable to you.

"I read every woman's magazine and shelter magazine every month," my friend told me. "I'm interested in what new trends seem to be forming, for one thing. For another thing, I look at all the ads very carefully. Advertising creative directors and art directors who do those ads supply me with a lot of ideas for new directions to be thinking about. I don't look so much at the product these people are featuring, but rather I look at the props and decorative details they use in their layouts. These people are paid to be innovative, creative, and appealing to the consumer. So I look for the little details they use to establish some interest with their consumer audience. More often than not, those details can translate into either new product directions or improvements. Because of my interest in housewares, I read the food ads carefully. I also look at a lot of television for the same reason. For a long time I didn't even own a set until I realized that it was the best way to stay current with the new trends. Watching TV commercials is now a part of my job.

"I buy and read every new kind of cookbook that comes out," my friend continued. "That's how we got into the fondue pot business. We were the first mass merchant to offer a fondue pot in a catalogue. When I first listed them, they were part of a Portuguese copper line, and I thought they were some kind of chafing dishes and said so to our customers. But by reading a cookbook, I learned what fondue was and that you needed a fondue pot for it. It was good business for us for years."

The same kind of thinking our buyer uses to recognize new ideas can be used by you to come up with new ideas.

There's a human factor to ideas that even some of the big companies recognize. A vice president of a major international toy company spent time explaining to me about their 250-man development groups, but then he admitted, "The key to development of any one, single product usually is one person. It's the one person who is really pushing for it. He may have a team of ten or so people working for him, but it's usually one person's drive that makes it happen." In other words, even at one of the gigantic companies that dominate an industry, it's one key individual who makes a new product happen. You, sitting in your kitchen, can do the same thing that individual does. In fact, if there's a lesson to be learned from all the big companies, it's this: You, working alone, with limited resources, can do very nearly anything they can do. You can do their kind of thinking. You can come up with the same product they can, and you can market it.

ANYBODY CAN be an INVENTOR

As long as we're on the subject of the human factor in the business of inventions, there's a fascinating story my buyer friend told me about some research that was done recently. Instead of researching products and public reaction to products, this time the mind probers were working on something more basic. One of the best, most innovative people at a major manufacturer was selected by his company to create a test that would isolate individuals with intuitive imagination. They wanted a test that would tell them whether someone was likely to be a good inventor or not.

After working intensively for three years with some of the finest psychiatrists, psychologists, and behavioral scientists in the country, they concluded it was not possible to isolate such individuals. There was simply no neatly delineated, easily recognizable pattern to good inventors. Which is simply a very costly, highly sophisticated, carefully measured way of saying what I told you earlier: Anybody can be an inventor.

For his own part, my buyer friend has devised a simpler way of finding good inventors. He likes to find people who have had door-to-door selling success—people who have made their living that way for at least three years. He believes you'll find certain basic characteristics in those persons that are also found in most successful inventors. Not the least of those characteristics is perseverance. A super-successful door-to-door salesman knows that he can sell three out of every ten homes he calls on, if he's really cooking. More importantly, he knows he *will not* make a sale at seven out of the ten. He's aware that he'll fail fully 70 percent of the time. He is therefore prepared to spend a lot of time being told "no" for the few times he hears the magical word "yes."

That kind of thinking reflects a high degree of intellectual fortitude. It takes a special kind of courage to be turned down time after time and keep going back for those three out of ten who won't turn you down.

Finally, a door-to-door salesman is an intruder. He knows it, he can deal with it. So is an inventor an intruder. The inventor constantly finds himself intruding on some manufacturer who's really too busy to be listening to yet another idea. (A vice president of creative services at a major gift company gave me a beautiful image of what happens to someone in his position when he's listening to somebody's great new idea. As he says, the inventor has to understand that the guy he's trying to sell his idea to is sitting there behind his desk with a great big thought balloon over his head like in the cartoons. In the balloon is one word: Risk. The inventor is barging in on the man, trying to shake up the order of things, offering him a swell chance to take a Big Risk. No wonder it's hard to sell ideas.) The inventor also finds himself intruding on the marketplace, which is already glutted with products, new and old. And he's intruding on the consumer who is reasonably satisfied with things the way they are and really not that interested in having to change his world around to accommodate a new product, even a good one that will help him.

Mr. Veep pointed out another reason an inventor needs intellectual fortitude. The poor inventor may have a marvelous idea he's trying to sell and still may strike out time after time. The fact that the idea is rejected doesn't mean it's a bad idea. It may just mean that the idea doesn't fit the company's particular needs. The inventor has to have enough courage and faith in his idea to keep trying until he finds a company that a) recognizes the worth of his invention (or idea), and b) needs that specific invention (or idea) for its product line. A successful idea really is one that meets two interlocking needs: a public need for that product and a company's need for that kind of product in the line.

THE EGGS BROKE

Every once in a while, there will be a great idea that meets all the requirements, and it still flunks. My VP friend gave a classic example that happened at his company a few years ago. A lady had made a hobby of hand painting incredibly intricate and lovely designs on eggs—sort of like the elaborate Russian Easter eggs of a century ago. One day she showed up with some of her hand-painted eggs and suggested that they might make beautiful illustrations for a line of Easter cards or, perhaps, a book on the art of egg designing. The eggs really were lovely. And he *could* use a nice new line of Easter cards. So my friend sent a professional photographer out to her home to take photos of the eggs.

The results were frustrating. Somehow photography was unable to capture the beauty of the eggs. It simply didn't translate from the three-dimensional work of art to the two-dimensional photograph. After a year of trying, everybody involved finally had to conclude there was simply no way to make it work. They had to turn down the Egg Lady despite the fact that the idea met a need both in the market and in their line.

When you're thinking about ideas, by the way, don't succumb to the traditional inventor's syndrome of "oh, well, somebody must have thought of that already." In the first place, you ought to make it your business to find out if it's been thought of already. (Part of what happens when you go to get a patent involves finding an answer to that question, by the way.)

Anyway, if it's not on the market, it probably never has been thought of. Even a bad idea, if it's backed by enough determination, will show up on the market at least briefly. As someone who's interested in new inventions and ideas, you'll probably notice it. Even ideas that have been thought of before may still be viable if you've got some new angle. A major toy executive noted, "We start a lot of projects that are subsequently abandoned for one reason or another. However, someone still could come in with basically the same idea but with some small differences, a fresh approach, and make the sale to us. It's worth it to make a royalty agreement," he added, "rather than try to fight any legal battles with an inventor." Even though the manufacturer may have solid proof that it worked with the same idea much earlier, that major toy company, and most other big companies, would rather pay the inventor and get a clear title to the new or newly modified idea.

THINK HORIZONTALLY, too

That also brings up another point. When you're having your idea, try to develop it horizontally as well as vertically. In other words, don't just work out all the details of the invention for a single application. Try to think of different directions to go with your basic idea, different models, different styling, and so forth. For example, if you're working on some new kind of design for a table lamp and you're using a floral motif, work out some other motifs too. Why? Because the guy you try to sell your idea to may be totally turned off by flowers but really hot for zodiac figures. That's kind of a dumb example, but something as dumb as that has been known to make the difference between an idea sold and an idea rejected.

Horizontal development means thinking of other applications for your idea, too. When you've got the basic idea in mind, spend some time thinking of *all* the different ways it could be used. Even silly ones. It could lead you into a whole new marketing direction in addition to the one you were already thinking about.

Whatever you do, don't stop trying to have ideas. Even if you've got a good one that you're working on, keep thinking of more and don't lose good ones that spring to mind at odd moments. I keep something I call (rather inelegantly) my "creative crap file." It's nothing more than a collection of odd bits of paper on which I've scribbled down ideas that have occurred to me and seem to have some promise. If I don't write them down or sketch them when they come to me, I'll lose them. At the moment I'd guess that there are more than two hundred reasonably viable ideas in the file, waiting to be developed when I can fit them into the schedule with the stuff that's already in the works.

With luck, you can start a good file of creative crap, too. I say "with luck" because I think anybody who is in the idea business is fortunate. Inventing and marketing new products is one of the more challenging, rewarding ways I can imagine making a living. I enjoy it immensely and I'd like to see you share some of the enjoyment. There's really no way of knowing exactly how many of us inventors and potential inventors are out there. The nearest thing to a statistic is the fact that last year 169,154 new patents were issued. Of these, over 85,000 were issued to U.S. resident inventors and over 20,000 were issued to U.S. independent inventors.

Next year, I hope you're one of them.

What's
A Good
Idea?

5

Some great ideas are born losers.
How do you tell if yours is
one of them?

IF YOU WEREN'T ALREADY INTERESTED IN INVENTIONS, YOU WOULDN'T be reading this book. After all, no matter how easy I try to make it, learning the ins and outs of becoming a successful inventor isn't exactly light entertainment. You could have more fun watching a grease job.

So, since you're here instead of watching TV or catching a good movie, I can only assume you're pretty serious about this whole business of making money with your idea. Good for you! That means you're ready to start thinking about the first hurdle on your way to success—figuring out just how good this idea of yours really is.

It doesn't count that your Uncle Ernie thinks it's the greatest thing since the invention of the inside straight. It's not enough that your neighbor says everybody would want one. Your uncle and your neighbor—wonderful persons though they may be—are a little less than totally objective about your invention. You need something a little more concrete to go on before you sink the family fortune, or at least a lot of the family's time and effort, into your idea or invention.

The question at hand, then, is the one at the top of this chapter: "What's a good idea?" I'll give you a short answer.

I don't know.

Nobody does.

There simply is no surefire way of telling if you have a good idea on your hands. Believe me, if I knew a foolproof way of finding that out, I would have retired a long, long time ago. I also would have saved myself all kinds of trouble down through the years because I've spent plenty of time and money chasing what turned out to be bad ideas. Anybody in the idea business who says he hasn't done that at least once is probably lying at least a little.

All is not lost, however. You may not be able to tell a good idea when you've got it, but a *loser* you can tell. No guesswork and gambling about it. There are some reasonably concrete requirements for a good idea, and if your invention doesn't meet those requirements, it's a loser. I still can't guarantee that your idea will be a success if it meets all the requirements, of course. All I can do is give you a way to tell whether you've got a turkey on your hands.

No, there isn't any nice, simple rule that you can write down on a piece of paper and carry around in your wallet. The key is a little more complicated. It is an understanding of how business works.

THE THREE-CUSHION SHOT

It isn't enough that your product seems desirable to a lot of people. Nor is it enough that the price seems right. It has to be both of those things at the same time—and one thing more. It must be profitable to the seller, the manufacturer, and all the other various people along the line of distribution. Every successful product is a three-cushion shot that touches all of these points: desirability, value, and profitability. If somebody offers you prunes at a good price, but what you really wanted was plums, there's no sale. On the other hand, if somebody offers you the plums you want but at a $100 a pound, there's no sale either. And unless everybody from the farmer to the grocer can make a fair profit out of the transaction, nobody's going to offer you either one—plums or prunes.

Let's take a look at the first of the three points, desirability. It's the toughest, most subjective of all the three points you have to meet. Not that the definition is difficult. If anything, it's simplistic: A product is desirable if a market exists for it. But simplistic or not, big companies spend millions of dollars every year trying to see if their products fit that magic definition. To give you a few examples of the kind of testing that goes on:

To update this book, I visited with a vice president of one of the giant toy companies in America. Every year, it adds approximately twenty new items to its line. In other words, twenty times per year, that company has to go through all the steps of developing and marketing a new product. Those twenty winners are selected from among some two thousand new ideas reviewed every year. So the early discards are pretty much on the basis of their own feelings about the market. (When I say "they," I'm referring to the triumvirate at the company in charge of selecting and developing new ideas: the president, the vice president for new products, and vice president for marketing. They work with a twelve-person staff.) Previous experience with a similar product, feedback from their retailers or sales organizations, analysis of market trends, and sometimes just gut instinct honed by many years in the business tell them to drop this idea or that one. Eventually they narrow the field down to twenty ideas that they believe will be winners.

Do they just dive right into production, distribution, and merchandising? Hardly. The next step is to mount a full-bore research project on each of the twenty ideas. No matter *how* sure they are of a product, nothing goes to market without some form of testing. There's just too much at stake. Basically, the research they do falls into one of three different levels of sophistication.

IT TAKES all KINDS

With ideas they're pretty sure of, they'll do a fairly unsophisticated kind of testing. They have artists make up some well-finished drawings or "comps" (short for "comprehensive drawings") of the products. They'll show these artists' renderings to a few hundred people who seem to fit the profile of their market—the right number of children, the right income, education, and so forth. And they'll interview test subjects about the product. Would they buy it? How do they think their children would like it? What would they consider a fair price to pay? Things like that. Perhaps no more than a dozen or so questions, depending on the kind of information they are seeking.

The next level of sophistication is basically the same kind of testing, with one important difference. In this level of testing, Mr. Giant U.S. Toy Company will actually make a model of the new toy. The model will be practically identical to the way the toy will look when it's manufactured. These models will be photographed—perhaps even with children. Now the interviews are conducted with these photographs of the product instead of the artists' renderings. The test subjects need to use less of their imagination to figure out what the toy will be like, so their reactions are likely to be more accurate reflections of how they'll feel about the real toy. Again, their answers to the test questions will be carefully analyzed to get an idea of how popular the toy will be, to determine what the price level ought to be, and to ferret out any unnoticed drawbacks about the toy. Naturally, it's more expensive to make a model and have it photographed than to just do a drawing, so this level of sophistication costs a bit more than the lower level.

The most sophisticated type of testing is reserved for toys about which Mr. Giant U.S. Toy Company has some doubts. (Obviously, they couldn't have any really serious doubts or the toy wouldn't have made it this far.) For this type of research, the toys will actually be produced as prototypes, either handmade models or production items from a short test run in the factory. They'll be as much like the real thing as possible—so much so that you and I probably couldn't tell the difference. These samples will then be given out to groups of children who have been brought by their parents to the test labs. While the kids play with the toys, the researchers will watch their reactions through one-way mirrors, which is fun. Later they'll probe the parents' reactions, too.

In the process of finding out whether a market exists for any given toy, the biggies regularly spend as much as $250,000 for research on an item. Sometimes they can get by more cheaply, but if they think they need it, they'll spend the top dollar just to be sure. They could lose many times that much if they make a million of the toy and wind up selling only a hundred thousand.

At another of the giants in the toy industry, they sort of back into the question of desirability. The first stage in development of a new idea there is to look at some segment of the toy business and select a price level for a new toy to be marketed in that segment. A typical starting point, according to a vice president, might be a suggestion from the marketing department saying, "We need for year after next a twenty-dollar doll with lots of 'action' features." (Note that lead time. If you try to sell your idea to a giant company, you'll learn that "next Christmas" is at least two years off.)

In other words, they look for holes in the toy market and seek to fill them with products. Work doesn't begin until the need for some specific type of toy is established. The idea people keep current on new materials and novel mechanisms, and at weekly conferences they attempt to fit these new things into ideas that fill the current product need. At these meetings, the marketing people play a very important part. Even before a new toy is beyond the handwaving, sketch-making stage, the marketing experts are working out the TV-promotion possibilities, the advertising and packaging ideas, and other related possibilities of the toy.

As you can see, marketing is the name of the game. And it seems to work.

JUST LIKE the PROS

It's worth pointing out at this stage that you can do some of the same sort of testing. Almost all market research is aimed at finding out to whom the product appeals, how appealing it is, and whether or not the price is right. No matter how sophisticated the techniques used to find the answers, the questions are usually variations on those same themes. Those are all questions you can ask, too.

There are several ways you can ask the questions. For one, you might simply do a short questionnaire requiring only yes/no or one-word answers, then buttonhole people at a shopping center and ask them your questions.

You might want to do an "in-depth interview" kind of study where you get a group of people together and steer the discussion into areas you're interested in. (This is more difficult, since you'll want to tape record and transcribe the whole discussion so you can analyze it all later.) Needless to say, you want your panel to be unbiased strangers, not friends who'll tell you what you want to hear. There are even a couple of variations on this "in-depth" research: (1) you might have a picture or a sample of the product for them to discuss, or (2) you might simply get the discussion rolling on the general market your product fits into, without showing or bringing up a specific product.

The advantage of the "in-depth" research over the questionnaire study is that you can turn up problems you didn't think of and learn product advantages you hadn't thought about. The disadvantage is that this is a highly subjective kind of study, and unless you're pretty well-grounded in market research, you could jump to some unjustified conclusions. If you use it, take the comments at face value and don't try to read more into them than that.

IF in DOUBT, TEST

The questionnaire study is much simpler and less likely to mislead you. Here's an example of the kind of questionnaire you might use:

1. Do you like the product?
2. Do you think that price is about correct?
3. What feature do you like most about it?
4. What feature do you like least about it?
5. Any comments?

You'll also need to know who it is you're talking to. You don't care about names and addresses, of course. And people are more likely to answer your questions if they feel they are anonymous. But you do need to know some basic information about them so you'll want to ask these questions:

1. Age (Check one)
    - ☐ 18 to 24
    - ☐ 25 to 39
    - ☐ 40 to 65
    - ☐ Over 65

2. Income. (Check one)
    - ☐ $20,000 to $40,000
    - ☐ $40,000 to $60,000
    - ☐ Over $60,000

3. Education. (Check one)
    - ☐ 8 years or less
    - ☐ 9 to 12 years
    - ☐ 13 to 16 years
    - ☐ More than 16 years

4. Male ☐ or female ☐.

5. Other questions pertinent to your product, such as marital status, whether or not the respondent has children, and so forth.

Keep the questionnaire short. The shorter it is, the more willing people are to answer your questions and the more interviews you can complete in a given time. Avoid questions that require long, narrative answers. Ideally, you'll have a sample of the product on hand. Otherwise, you'll have to make do with a picture of it. And if you have a sample, be prepared to replace it from time to time after it is dropped or broken by all the handling.

Get as many interviews as you can because that cuts down on the amount of error introduced by a couple of nontypical responses. And be sure you hit more than one location when you are doing your research. (If you do all of it at one shopping center, or even at a couple of places in the same general part of town, you could build in an unnecessary bias in your sample. You could wind up with a heavy sample of well-educated, wealthy respondents, for instance. Or the opposite.)

Your specific product may dictate special variations on the questions, of course. But you'll have to work that out on your own. Just remember that the object of the research is to give you insights into how well received your product is likely to be. It may take some time and effort and involve a certain amount of expense, but it could pay off handsome dividends if it uncovers some problems you didn't know about.

If you're systematic about your interview and keep careful records of the answers, you could have a valuable sales tool to use later on when you try to sell your idea to some manufacturer or when you try to raise some financial backing. Careful research, properly handled, can be very, very impressive. Especially to investors.

AN ABSOLUTE PERHAPS

While we're on the subject of research, I have to point out that it is hardly infallible. "Market research is not a science, by any means," said a vice president for new products at that biggie toy company that I mentioned earlier. "At best it can sort out the extremes. It can't tell you ahead of time whether to expect 200,000 sales or 800,000 sales. All it can do is tell you you'll probably do somewhere in that area."

There have also been some monumental flops in the research business. The late Seymour Popiel, inventor of the Vegomatic, the Pocket Fisherman, and a number of other hugely successful products was the last person in the world I'd expect to have been stung by bad research results. Seymour was not only the master of his very special kind of marketing—TV promos with their high advertising budgets and heavy promotions—but he invented it. He carried the midway pitchman out of the carnival and put him on television, with fantastic success. He made research a cornerstone of that success, developing it to a very fine art in all his promotions.

Anyway, Seymour—like the rest of us—noticed the introduction of a new product called the trash compactor a few years ago. Brought out by Sears and Whirlpool, jointly, it was making heavy inroads in the kitchen-appliance market. So when Roy Bradbury II, an inventor, called on him one day, Seymour was ready to listen to the man's description of a nonelectrical, low-priced trash compactor. It was a fascinating concept. The basis of the unit was a kitchen stool with a cylindrical base. You put the trash and garbage in the cylindrical part, then put the stool down on it and you scrunched the garbage by sitting on it or standing on it.

It looked like a great idea to the idea expert. But part of Seymour's success was due to the fact that he never—but *never*—marketed anything without doing a lot of research. He made samples and had them tested by experts. The garbage scruncher, which was called the "Sit On Trash Compactor," was probed by consumer panels, home economists, engineers, and marketing experts. It was restyled three times before it passed muster, but finally all the tests were positive. It was going to be a winner.

Seymour produced his usual batch of hard-selling television commercials, manufactured 65,000 of the new gadgets, put the commercials on the air at $19.95 and $29.95 in different markets and stood back to make room for the thundering head of buyers.

There was a deafening silence. Nobody bought the things. To this day, no one really understands what went wrong except that somehow the research produced a clinker. Seymour, eternally the pragmatist, didn't bother to spend another fortune finding out how come he lost all that money.

Since including this little vignette in my first version of this book, I learned an interesting fact indirectly from the inventor; it seems at the very moment Seymour chose to introduce the Sit On Trash Compactor, the price of plastic soared to a point where Popiel was not sure his desired profit ratios could be maintained even were the scruncher a solid hit.

So, Mr. Popiel closed out the remaining garbage stools at $12.99 each, washed his hands of the whole thing and moved on to bigger and better things. Rest in peace, Seymour.

That biggie toy company had a big research failure in reverse a while ago. They brought out a toy called the "Funny Farm"—basically a hundred-year-old idea in modern dress. It was a toy farm, with buildings and equipment and livestock and people all done in plastic. The price was pegged at about $14 or $15 at retail and conventional wisdom said it wouldn't be much of a seller at that price. The rule of thumb in those days was that any toy with a price tag over $10 would probably be a slow mover. Research results confirmed it. The best Mr. Big figured on selling was maybe 250,000 units a year.

What the research didn't show—*couldn't* show—was that the year the toy came out, toy dealers around the country were all looking for a good value to promote. A sort of "loss leader" they could advertise to bring people into the stores. It had to be something that represented a great value for the money and that would be a popular toy. The Funny Farm set fit their needs exactly. The toy dealers could afford to sell it for $9.95 (just under the magic figure of conventional wisdom), and at that price they wouldn't get hurt. It was obviously a good deal, and so thousands of dealers stocked up heavily and sold out over and over again. Instead of leveling off at 250,000 sales, as the research had predicted, Funny Farm took off like a rocket and sold something like 700,000 units that first year. Big Toy Co., Inc. had to scramble like mad to keep up with reorders, since they hadn't geared production for that kind of demand.

TINKER, TINKER, LITTLE STAR

Every now and then professional research will get all tangled up in its own feet, too. At Sears, the product-design staff went to work on a pecan sheller brought in by a doctor from Georgia. The doctor had forty acres of pecan trees and had been selling pecans for some years as a sort of cash-crop hobby. A tinkerer by inclination, the good doctor devised a little gadget that shelled the pecans mechanically. He used it on his own pecans because shelled nuts always bring a much higher price than unshelled ones. But it occurred to him that housewives could use the sheller to save money. They could buy the cheaper pecans, still in the shell, take them home, and turn them into expensive shelled pecans with his gadget. It made a lot of sense, so the doctor set up a small factory in an old post-office building near his home and began turning out the shellers. It was a pretty simple device, with a couple of moving parts and some rubber bands. It looked a little crude, but it sure did the job. Sales grew, and the doctor turned the sales and distribution over to a sales outfit in the South where it began doing a brisk business. Finally, it became such a success that the doctor and his pecan sheller were discovered by Sears!

Sears liked the idea, but it felt the gadget was too crude looking for its market. So Sears assigned a top industrial designer to streamline the doctor's design—to get rid of those hokey, old rubber bands and generally do a cosmetic job on the gadget. Before it was over, five engineers had taken a shot at improving the doctor's design, and they all gave up. With a sigh, Sears put it into its retail distribution in the sheller's down-home, funky, original form and proceeded to sell it at the rate of $1 million a year.

No, that's not marketing research, so it's really a slightly different category than what we've been talking about. But here's the interesting sidelight—my Sears buyer friend took a gross of the shellers to a church bazaar to demonstrate it and see what kind of sales results he could get with it. He wanted to get some idea of how it might do in the catalog Sears used to publish. My friend was an old door-to-door salesman, and he was no slouch when it came to something like this. If anybody could sell that sheller, he was the man. He rolled up his sleeves, shelled the hell out of a bunch of pecans, and worked the crowd like an experienced pitchman. Sales results: zero. Nobody bought it. He went home with a sinking felling that he had a real loser on his hands, that the sheller was going to *die* in his stores.

As we know, of course, it sold beautifully. He said, "What if that sheller had been professionally market tested before we ever offered it for sale? It would have failed before it ever had a chance to succeed. We would not have put it in the catalog, and we would have missed $1 million in sales per year."

EVOLUTIONARY IDEAS

The point of all this is that research can be fallible by a long shot. The research can err either way —it can give false indications, and at best it can only sort out extremes. But for all of that, it's one of the few ways of measuring desirability. Desirability remains a pretty subjective question, no matter how much research you do, so it's nice to know that there are other things you can do to make sure you have a desirable product in mind.

For instance, nearly everyone I talked to in doing my research for this book told me the same thing. "We're not looking for something completely new, no total breaks with what's gone before. Rather, we're more interested in finding a variation on an existing theme, an *evolutionary* rather than a *revolutionary* development."

A vice president of new products at a major toy company told me, "Ninety percent of all new ideas really consist of taking an old, basic idea and using new materials and styling." (For example, the Funny Farm toy that took off so unexpectedly.)

The Barbie Doll at Mattel is another classic example of that kind of product. Dolls you could dress up have been around for years. Barbie simply translated that idea into an industry by being the sexiest doll ever thunk up. The kids enjoy it more because it is more realistic to put a real dress on a (wow!) three-dimensional doll. Mattel enhanced that realism by giving Barbie a boyfriend, a family, a choice of careers and all the accoutrements of the modern "good life" as seen on TV. And, of course, Mattel enjoyed it more because it put paper dolls into a new, higher-price category and created a complete aftermarket. (Having bought the doll, the kids would want to keep adding to the wardrobe and the collection of status symbols for Barbie, Ken, et al.) It was an evolutionary development, but it revolutionized toy marketing.

The Die-Hard battery at Sears & Roebuck is also a great example of evolutionary ideas being winners. The major difference about the Die-Hard and other batteries is that the walls of the battery are much thinner and lighter than those of previous storage batteries. More of the battery is the part that makes electricity, instead of the part that holds the battery acid in.

Thus, you can get the same amount of power out of a smaller battery. Or you can get more power out of the same size. I imagine that the Die-Hard battery is the most successful invention of its kind in history. Sears must have sold a jillion of them.

So when you're thinking about your invention, see if it emulates the Barbie Doll and the Die-Hard battery approach. If so, you have made a good start. As that VP told me, "A successful idea needs to have some relevance and continuity with the past. The new idea ought to be the next logical step in a progression." In other words, evolutionize, don't revolutionize.

Another way to ensure the desirability of your invention is something we touched on in the last chapter. Make sure your idea involves an existing problem. Logically enough, if the problem you solve is a common one, and the solution you devise is a really good one, you've got a desirable invention. Measure your idea against that benchmark!

There's one final aspect of product desirability that you ought to bear in mind. You may have a perfectly desirable invention, but if it is only desirable to a very limited group of people, it may not be marketable. If the only people who can use your invention are left-handed lady pinochle players, you're going to have a devil of a time getting it on the market. That's just too limited an audience to bother with in terms of profit. There's not enough money to be made in selling it.

Anyway, whether you use some form of research or testing, whether you try to think of problems to solve, however you go about it, the first consideration is whether or not your invention will be wanted by a sizeable number of people. Once you're past that hurdle, you have to figure out if it can be made for the right price.

HOW MUCH is MUCH?

The exact steps you have to take in determining the selling price of your invention are covered in some detail in chapter 8, "Doing Your Homework." For the moment, the point is that the price has a lot to do with the success or failure of a new invention, no matter how many people are willing to pay to own your product. And that's not quite as straightforward a proposition as you might think.

The price acceptability of a product actually varies, depending on where it's sold. You know yourself that you might see some product in a discount drugstore and be shocked by the price—but not bat an eyelash at seeing the same price in a swankier store. For that very reason, many manufacturers make the same product with a different brand name, different packaging and—of course—different price tag. The upper-crust brand name will be sold at Madame Fifi's Giftie Boutique, while the plebian version will be hanging on an endcap at Cheapo Charlie's Discount Mart. They'll both sell like gangbusters, too. Same Product. Different package and price.

Price acceptability works the other way around, too. It is actually possible to make a product sell too cheaply. By doing so you obviously cheat yourself out of some profit. But what's less obvious is that you could also cheat yourself out of some sales. Your invention might be highly desirable, but if the price is too low, people may be unwilling to buy it. They may have qualms about how well made it is, or they may have feelings that it is somehow beneath them to buy something that inexpensive to do what your invention does.

Ultimately, the only way to find out if your price is right—once you've done your homework and calculated what you'll have to get for the product to make a profit—is to do some research. Check the stores to see what similar or related products are going for. Find out how much price variation is going on in the field your invention fits into. See whether higher- or lower-priced versions of products in your field are being marketed in different kinds of stores. In other words, get a fix on what the price range is for products like yours and figure out if you can play in that league. And if you do some of that supermarket-parking-lot research we were talking about a while ago, be sure you ask some questions about price.

Finally, there's the question of whether your product will make a profit for all the people along the line of distribution—manufacturer, sales organization, wholesaler, retailer. Again, it's not as simple a question as you might imagine. What you consider fair and reasonable profit for a manufacturer might not be fair and reasonable at all to the manufacturer with whom you're dealing.

Suppose, for instance, you're hoping to make a royalty deal with a manufacturer of automotive accessories. You've calculated your projected retail price on the basis of allowing 50-percent markup for the manufacturer. That might sound just fine to you. But it's possible your manufacturer only works on items with a 100-percent markup. Your deal isn't going to interest that manufacturer.

What it boils down to is that you're going to have to do a little research into the markup structures involved in the industry where your invention fits. Needless to say, the profit structure along the way from manufacturer to retailer will have an effect on the price of your product, which will have—as I said—a lot to do with whether you have a bad idea or not.

If you find that your invention covers all three main bases—desirability, value, and profitability—there's still no guarantee that you're going to make a lot of money with your new idea. As I pointed out earlier, there's just no way to tell whether your idea is a winner or not. All you can tell is whether or not it's a loser.

Patent
IT, Fast!

*Getting legal protection
for your idea.
What it costs, what it does,
how to do it.*

SINCE I AM NOT AN ATTORNEY, I FELT IT WOULD BE APPROPRIATE TO have Scott Kelley, my patent attorney for over seventeen years, review this chapter. He did this for me, and I wanted to thank him right up front for his effort. Scott is, incidentally, a partner in the Law Offices of Kelly, Bauersfeld, Lowry & Kelley, LLP. Their offices are in Woodland Hills, California. If that's in your neighborhood, be assured you would be well protected were you to retain Scott's group for your Intellectual Property needs. At my request, and at his direction, Scott's secretary has developed a list of well-recommended patent attorneys across the country. Were you to call her at (818) 347-7900, she'd be happy to fax you a list of qualified patent attorneys in your area. Now, it's time to get back to my business.

EVERYBODY would WANT one

I was standing in a line not long ago, and to pass the time I struck up a conversation with a stranger. She was a motherly looking woman, probably hard working, certainly not very wealthy. And it turned out she had an idea for an invention. An idea that was going to put her on easy street.

She had no way of knowing that I was writing a book about that very subject, and I didn't tell her. I wanted to see if the thing I expected to happen would happen.

It happened. It always does.

The conversation went something like this:

"It's really a great idea," she said.

(So far, so good. I've never yet heard anyone say, "I've got this mediocre idea here . . .")

"It wouldn't cost much to buy, and just about everybody would want one," she continued.

"What sort of invention is it?" I asked.

"Oh," she said, embarrassed. "I wouldn't want to talk about it yet. Not until I've sold the idea. Someone might steal it."

"Have you thought about getting a patent for the idea?" I pressed. "Then you'd be protected."

"Oh, no. They cost so-o-o much money, and my brother-in-law told me I don't need one. He said all I have to do is mail a registered letter to myself explaining the idea. That gives me proof that I had the idea first, in case anyone tries to steal it and claim they thought of it."

And that was that. I never did find out what the idea was, of course, because she was convinced I'd steal it and run off to Acapulco with the fortune that rightfully belonged to her. I wonder how she plans to sell the idea to someone if she can't talk about it.

The point is that, like most people with an idea for an invention, she was terrified of having the idea stolen and equally terrified of getting the protection of a patent.

The important lesson you should learn now is this: The first thing—the very first thing—you should do with the invention you want to develop is apply for a patent.

WELCOME to PARANOIDVILLE

Hardly anyone does, unfortunately. Like the lady in the line, most first-time inventors are full of misinformation and trepidations about getting legal protection for their ideas. They live in mortal fear of having their idea stolen. But they dread even more the thought of getting a patent. All the money they'd have to spend, all the legal rigmarole they'd have to go through! You may feel the same way yourself. So, let's set a few things straight about patents:

44

* Just about everything you think you know about patents is probably wrong.

* Patents, while not exactly bargain-basement stuff, can cost a lot less than you think. They're certainly much easier to obtain than most people have been led to believe.

* A patent is the *only* real protection there is for an invention.

You can forget that old gag about mailing yourself a registered letter. The letter-to-yourself trick does offer some protection, it's true, but it is only partial protection at best. At worst, it's a trap. If all you have is a letter to yourself proving that you had the idea first, someone else still can take your idea and run with it. Worse still, he may even be able to get a patent on your idea! And you won't be able to stop him with your letter.

Why? Well, in the eyes of the law, there are two steps involved in the act of inventing something. First there is conception—having the idea. By mailing yourself the registered letter, you are establishing that you had the idea as of the date you mailed the letter. And as far as it goes, that's fine. It works. It's proof, all right, but it misses the crucial second step. The second step is called "reduction to practice"—putting your abstract idea into some concrete form. Your letter to yourself doesn't have a thing to do with this step. In fact, there are only two ways to reduce to practice: (1) You can make a model or sample of the invention. This is called "actual reduction." (2) You can apply for a patent. This is called "constructive reduction."

Okay. So there you are with your letter. And you get word that some no-good creep has stolen your idea and has applied for a patent. If all you have is that registered letter, your chances of beating him aren't worth much. In all likelihood he gets the patent and you get the short end of the stick. Even if you actually made a model—that is, you achieved "actual reduction to practice"—you still have to prove to the satisfaction of the government that you did it before he did. You've gotten yourself hip deep into a messy, costly, time-consuming legal battle. And you could lose.

Suppose the other guy hasn't applied for a patent either. Suppose he's simply gone into business cranking out your invention and selling it. "Aha!" you say. "I'll soon put a stop to this by taking him to court and showing the judge my letter, proving I had the idea first." Wrong. It doesn't matter who had the idea first if you don't have a patent. Only a patent will stop him from producing your invention and marketing it. Only a patent will prevent him from selling it to some manufacturer for a royalty deal. You can't touch him with anything short of a patent, and that's what you'll have to get before you can shut down his machinery. In the meantime, he can sell as many as he wants, perfectly legally. He may even completely glut the market with your product before you can get him stopped. And there you'll sit—no market left for your product, wishing you'd applied for a patent before things got out of hand.

There's one more catch to the registered-letter gimmick. It is a piece of evidence. Maybe your *only* piece of evidence. But shortly before going to trial with it, the odds are that you'll destroy that evidence. That's right. The majority of people do. The reason you send the letter by registered mail is to have proof of the date it was sent and to have an official United States Postal Service seal on the letter flap. The seal proves that the letter hasn't been tampered with and that the contents are the same ones you originally mailed. So far, so good. Now it's getting close to your day in court. It probably has been more than a year since you mailed yourself the letter. Now panic and doubt set in. Your whole case is based on the contents of the envelope, and you begin to wonder if you really remember exactly what the letter says, all the details of the invention you described at the time. Are you sure? Nope. So what do you do?

You open the letter. You destroy the seal. You make the evidence worthless.

Sure you're smarter than that. But a surprising number of people aren't. I suppose the way to beat that particular problem is to make a copy of the letter when you write it, then clip it to the sealed envelope when it comes back to you in the mailbox. But it still doesn't solve the essential problem of getting real protection for your idea.

There is no shortcut to protection. Getting a patent is much easier and better than relying on some legal folk remedy like the registered letter to yourself. So let's take a look at what that real thing is, how you get it, what it costs, and how long it takes.

To begin with, there really are a number of different kinds of protection available. I've been referring to them all loosely (and somewhat inaccurately) by the name "patent." In fact, the five different kinds of protection you can get are discussed as follows.

COPYRIGHT

Copyrights are for works of artistic endeavor. Paintings, photographs, poems, sculptures, computer source code, and this book are all examples of things that can be and are copyrighted. A copyright is, in effect, protection for "artistic creation" and is quite narrow in scope as compared to a patent.

TRADEMARK

Here's Scott's textbook definition of trademarks: A "trademark is a word, design, slogan, color combination, sound, smell, or combination of any of the foregoing, used by a manufacturer or merchant to identify the merchant's goods and/or services and distinguish them from others." Trademarks include brand names identifying goods ("Coca Cola" for a soft drink), service marks identifying services ("Marriott" for hotel services), certification marks identifying goods or services meeting certain qualifications ("UL" for appliances meeting the safety standards for Underwriters Laboratories, Inc.), and collective marks identifying goods, services, or members of a collective organization ("AFL-CIO" for union locals). The same legal principles generally apply to all of these terms, often simply called "marks."

The important thing to remember is that a trademark need not be registered to be protected. Reserving a corporate name in one or more states is irrelevant to trademark rights. Trademark rights can be acquired only in one of two ways:

1. By actually using the mark on or in connection with goods or by displaying the mark in the sale or advertising of services

2. By filing a federal intent-to-use trademark application, declaring that the applicant has a "bona fide" intent to utilize a particular mark in connection with designated goods and/or services

Registration of a mark is not mandatory, and rights will be protected without it. An applicant for registration of a trademark in the U.S. Patent and Trademark Office will only receive a registration certificate once goods bearing the mark have been shipped in interstate or foreign commerce, or if services under the mark either are subject to federal regulation or have been rendered in more than one state, or alternatively if the owner is foreign and has registered the mark in its country of origin or has applied to register it there during the previous six months.

When issued, which normally takes about one year, a federal registration has substantial advantages:

* It is constructive notice of the registrant's claim of ownership, applicable nationwide to everyone subsequently adopting marks.

* It will be listed on search reports obtained by others.

* It is evidence, albeit rebuttable, of the registrant's exclusive ownership rights, shifting the burden of proof to anyone challenging those rights, and in some circumstances it can be conclusive evidence of those rights.

* It gives federal courts jurisdiction to hear infringement claims, counterfeit claims, and related claims of unfair competition under state law.

* It precludes states from requiring modifications in the display of the registered marks.

* It can be used as a basis for registration in some foreign countries.

* It can be recorded with the U.S. Customs Service to prevent importation of infringing foreign goods.

Use of a trademark registration notice before the mark actually has been registered is not only inappropriate but may also prevent the owner from obtaining relief against an infringer. The informal symbols TM (for trademarks) or SM (for service marks) are often used with unregistered marks to indicate a claim of common law trademark rights. After registration, one of the following trademark registration notices may be used: "Registered U.S. Patent and Trademark Office," "Reg. U.S. Pat. & TM Off.," or "®." Such notice is not mandatory, but under some circumstances its use is necessary to obtain damages from an infringer.

Trademark rights continue indefinitely as long as the mark is not abandoned and is properly used. A federal trademark registration is maintained by filing a declaration of use during the sixth year after its registration and by renewal every ten years as long as the mark is still in use in federally regulated commerce. The law provides that nonuse of a mark for two consecutive years is ordinarily considered abandonment, and the first subsequent user of the mark can claim exclusive trademark rights.

THE FINAL WORD on PATENTS

Design patents are directed to the "ornamental appearance" of an object and are useful in preventing competitors from imitating the unique visual appearance of an article of manufacture. Design patents have a term of fourteen years from the issue date, and in comparison with utility patents, are relatively inexpensive. Utility patents, on the other hand, are directed to articles of manufacture, machines, chemical compositions, and processes. The term of a utility patent is twenty years from the filing date.

To be patentable under the utility patent statutes, an invention must be "new, useful, and nonobvious." The patent application does not consist of a form filed with a governmental agency but rather a full description of the invention and background materials relating thereto sufficient to enable one of ordinary skill in the art to make and use the invention. A patent application, therefore, typically contains drawings, a discussion of the prior art, a detailed description of the preferred embodiment, and claims. The claims define the monopoly the inventor would like granted to him or her by the government in exchange for a full disclosure of the invention.

It is often desirable to conduct a preliminary patentability search prior to filing a utility patent application. Such patentability searches typically cost between $900 and $1,500 and take one month to complete. Even if you are unable to locate "prior art" that might have a significant bearing on the patentability of your invention, experience shows that you can often recover the cost of the search during prosecution of the application. This is possible because by having the search results in front of you and your lawyer during preparation of the claims, you are able to tailor the claims much more closely to the scope needed to avoid the prior art. Thus, the prosecution process in the Patent and Trademark Office will be shortened, which results in decreased prosecution fees.

Within the last several years, it has become possible to file "provisional" patent applications in place of a regular utility patent application, in order to secure an early filing date. Provisional applications may be filed in an informal format and do not require patent claims. The provisional application must be followed by a regular utility application within one year in order to claim the benefit of the provisional application filing date. It is often desirable to file a provisional application as a first step and later file a utility application if the invention warrants it.

One advantage of a provisional application is that the term of twenty years from the filing date will mark from the filing date of the utility application, while prior art will be defined by the filing date of the provisional application. On the other hand, for filing priority, related foreign applications will need to be filed within one year of the provisional application filing date.

A DUMB DECISION

Those, then, are the different kinds of legal protection available to you. You can use them in just about any combination, depending on the nature of your idea. For instance, on the Rickie Tickie Stickies, I copyrighted all the material on the package and the display items I designed to merchandise the flowers. And I got a design patent on that illegible typeface I designed for the trademark. In those days, in order to properly copyright my flower design, I was obligated to put a ©RTS on the face of each flower. Being a design purist, I hated the idea of ruining my product with that ©RTS. I did not put the © notice on each flower. That was probably the dumbest legal or design decision I ever made. I did *not* get a design patent, because I figured that it was too easy to beat, in the case of the flower. Any minor change in the design would be enough to beat that patent. As we'll see later, that may not have been the wisest decision I ever made, either. The theory is to give your product as many kinds of protection as possible.

But what, precisely, is it that these things do for you? How do they protect you? They're not permits. You don't need a permit from anybody to manufacture and market your idea, work of art, or brand name. What you *do* need is a way to prevent someone else from doing it and making money from your stroke of genius. That's what your legal protection does. It gives you the exclusive right to manufacture and market your idea or to license the manufacture and marketing of it to someone else. The key word is "exclusive." You control the ball game. You own the license. If you choose to let someone else in on the license, you can charge money for it, or you can charge a royalty on merchandise produced under it.

Getting the protection of a patent, a copyright, or a trademark registration ranges all the way from ridiculously easy to fairly complicated and time-consuming. The cost varies from peanuts to a moderately large sum of money. It depends on what you're looking for and how complicated your invention or idea is. Let's take a look at patents for openers.

GO to a PRO

The first thing to do when you decide you want a patent is, logically enough, to call a patent attorney and make an appointment. I'm lucky in that respect. I've known Scott Kelley, a highly regarded patent attorney, since I first became involved in marketing ideas and inventions. In your case, you probably don't have the name of a patent attorney in your address book, so you'll have to do a little scouting around. Your banker may have a name or two for you. If you have a lawyer of your own, he may know the name of a good patent attorney in town. Maybe even the Bar Association of your county or state could help you out. One way or another, find one and tell him you'd like to meet with him to discuss a patent application.

The initial meeting is practically guaranteed to be easy and painless. Most patent attorneys charge only a very nominal fee ($100-$150)—or sometime no fee at all—for the first brief discussion of your invention. It usually lasts for about half an hour. And that painless half an hour may save you enormous amounts of blood, toil, tears, sweat, and treasure.

I don't know how many times Scott has had to tell an inventor at that first meeting, "Sir, that's a fine idea you have there, but it's been done. The patent was issued several years ago." Or on other occasions, he's said something like, "I'm sorry, but there simply doesn't seem to be any way to patent that idea. It sounds workable and saleable, but it's an obvious variation of things that are already in general use."

The wording of the federal statute on patents is worth quoting in regard to that last one. The law says, "A patent may not be obtained . . . if the difference between the subject matter sought to be patented and the prior art are such that the subject matter as a whole would have been obvious, at the time the invention was made, to a person having ordinary skill in the art to which said subject matter pertained."

In other words, if it's obvious—if it's simply a new combination of old, well-known objects or a new use of an old, well-known object—it's not an invention, and it can't be patented. There are any number of good ideas that fall into this category. Your patent attorney can tell you right up front if yours is one of them. Naturally, that doesn't mean you can't market it. But you won't be protected by a patent if you do, and you can expect competition to crop up fairly quickly after you get to market with it. You may also have a very hard time selling an unprotected idea to a manufacturer.

Sometimes the patent attorney might even break the news, as gently as possible, that the idea is a loser and the inventor is probably wasting his time. You don't have to take the attorney's word for it, of course, but most patent attorneys worth their salt have seen enough inventors come and go to be worth listening to.

When you see the patent attorney, you should be prepared to give a complete and fairly detailed explanation of your invention. Until he understands fully what your invention is and what it does, he can't start action on your patent application. Your presentation doesn't have to be elaborate. It might be nice to have some sketches or maybe even a model of the invention, but if you can talk and flap your arms well enough to get the idea across, that's fine. (If you have drawings, by the way, they don't have to be professional quality. Even if they were, they'd have to be redrawn for the patent application. The Patent and Trademark office has its own carefully specified style of drawing—which I hate. They won't accept anything else. There are artists who specialize in those drawings, and your patent attorney will take care of all that for you.) In the case of a design patent, where the whole thing lies in what it looks like, you may have to have a sort of drawing or sketch to get the idea across to the attorney.

Once the attorney understands your invention, he'll open a file, or docket, which sets your patent application into motion officially. That may not sound like a big deal, but it has one important benefit. It is legal proof that, on such and such date, you had the idea for your invention. This is proof of conception, which will stand up in just about any court in the land. (For whatever it's worth, you can prove conception by explaining your idea to someone and getting him to sign a statement that you did so, that he understood it, and that the date was so and so. This is at least as good as the old registered-letter routine.)

A FOOL for a CLIENT

Having opened your file, the attorney's next step is to conduct a search. This is a very critical part of any patent application. What's involved is this: The attorney (or more likely an associate of his in Washington, D.C.) goes to the Patent and Trademark Office and carefully checks out all the patents that seem to have some bearing on your invention. Any patents of particular interest are copied and sent back to your attorney.

Actually, you can do the same search in many public libraries across the country. A number of cities have libraries that are official patent repositories. But doing the search in the library will take longer and will most certainly be less thorough than doing it at the Patent and Trademark Office in Washington. The cross-indexing system in the library patent repositories is much less complete than the system used in Washington. Doing it at the library is doing it the hard way. I would also not recommend using the Internet; there is just no way to guarantee that you've done a thorough search. This is a job for a professional.

The object of the search is to make sure that no one has already obtained a patent on your invention or something pretty close to your invention. It also uncovers other inventions that may be able to do some of the things your invention does, which will have a bearing on the claims you can make for your invention in the patent application.

You may be able to skip the whole search, by the way. If the invention is in a very specialized area, in an area which you are very familiar, you may not need a search. You'll know all the products on the market and in use in your specialized field, and you'll know, for a fact, that your idea is unique. This sort of situation is rather rare, but it *does* happen once in a while.

STAKE your CLAIMS

The first page or so of your patent application is pretty much spelled out by the law. You must offer a detailed description of the invention and some background on the area of interest to which it applies. There must be a description of what the law calls its "best mode of operation"—how to use it. And there must be drawings of the invention, except in cases where that's impossible. A chemical process, for instance. This whole section is called the "Specification."

The next section is a series of numbered paragraphs called "claims." This is where you put forth your notion of the unique, exclusive features of your invention. The features that make it unlike anything covered by an existing patent. Here is where there is almost always a lot of haggling back and forth between your attorney and the Patent and Trademark Office. The government is interested in keeping your claims as narrow as possible, limiting them to the most specific and least sweeping ones. You, of course, are interested in claiming as much as possible for your invention, and in the broadest possible terms so that you have more protection.

I'd better explain a little bit about broad versus narrow claims. Let's say, for the sake of illustration, that you're trying to patent your new process for turning base metals into gold. The patent office may well come back and say you can only claim to turn lead into gold, based on their interpretation of your specification. Back and forth go the arguments, volley and thunder, until some mutually acceptable compromise statement is agreed upon. If you have to settle for a patent limited to the process of turning lead into gold, someone else may be able to patent a process for turning brass into gold. If, however, you've proved to the satisfaction of the patent office that your process will turn *any* base metal into gold, you're covered, and the new guy loses. That's what I mean by narrow versus broad claims. The more you can cram under the umbrella of your patent, the less there is left over for someone else to patent.

In putting forth your claims on the patent application, you put the broadest claims first, getting narrower as you get farther down the list. Why bother putting in the narrow claims, too? Because later on in your negotiations with the Patent and Trademark Office it may turn out that your broad claim is not protectable. Then you can fall back on the narrower claim, and so on down the list.

The whole application—all the drawings, all the legal statements, and the patent search—will be handled by your patent attorney. All you have to do is sort of help out when he needs some additional information about your invention. And you have to pay the man.

DON'T be CHEAP

As you might expect, the cost of a patent application varies all over the lot, depending on how complicated it gets and how long it takes.

You can figure on spending something like $800 to $1,500 for what is called a "preliminary novelty search," which is a search to determine whether or not your idea is novel, or, in other words, new and unique. For a little bit less you can have a "state-of-the art" search that will tell you, for instance, which patents are in effect regarding vacuum cleaners. Or lawn mowers. Or whatever. This sort of search is helpful if you're trying to come up with some sort of new idea in a specific area. It's no help for someone who already has invented something and wants to find out if it can be patented.

Filing the application is the next cost. A utility patent application for a relatively simple idea—one that can be shown in one page of patent office drawings and a relatively short set of specifications and claims—will cost anywhere from $2,500 to $3,500. Unless you have a pretty technical invention, this is about what you could expect to pay. Naturally, filing the application for a plant patent or a design patent costs less since there is less work and time involved.

Then there's the cost of all that negotiation between your attorney and the patent office over your list of claims. Depending on how hot the fight waxes and what the attorney has to go through to win for you, the fee may go from $1,000 to $5,000. About $2,000 is a reasonable figure to work with.

IT'S SWEATY-PALMS TIME

As you can see, the real sweaty-palms period of the patent application is during the negotiation with the Patent and Trademark Office over your claims. Here's how that works.

After your patent attorney files the application, everybody sits back and waits nervously for the patent office to make its move. After a while, the patent office replies, usually saying something terribly discouraging. Like, "No." They may initially refuse to grant a patent at all, or they may grant only a very narrow claim. This letter is called an "office action," and it is just the opening gun in the battle that almost always ensues. Don't be distressed. Everybody (or just about everybody) has to go through this wrestling match. It's all part of the game.

The lawyer then sends back a thoughtful reply, pointing out to the patent office all the reasons you should be granted a patent after all. The patent office grudgingly reconsiders a few things, your attorney insists on a few more, and so the battle goes until somehow a resolution is reached. The patent office bases its objections on patents already existing and on published material that seems to indicate your idea isn't new. Your attorney, naturally, puts forth arguments that those other inventions aren't anything at all like yours.

One disappointing fact is that the patent office generally will *not* cite conflicts between your application and other applications which are pending. If there is another patent in the works that conflicts partly or totally with yours, you have no way of knowing. And the patent office won't tell you. They're not being mean or unreasonable, by the way. It's just that they can't very well talk to you about a conflict with a patent that hasn't been issued. Without a patent, there is no infringement. The first time anybody normally finds out about such a conflict is when one of the pending patents is issued, at long last. Then the applicant whose patent has not yet been issued gets a letter from the patent office. The letter points out that the applicant's invention now is in conflict with the other, brand-new patent.

Once in a great while two or more conflicting patents will near completion at almost the same time. When the patent officials see that kind of situation shaping up, they'll step in and tell all of the applicants involved that there is a problem. Each applicant then has the right to copy down all the claims on the other applications, study them and call for a federal "interference proceeding" aimed at parceling out claims among the conflicting patent applications in as equitable a fashion as possible.

Getting a patent takes time, of course. Starting from the beginning, the search usually goes on for about six weeks. Then your attorney needs a month or so to prepare the patent application, including all those specifications, claims, drawings, and so forth. Finally, there's all that haggling with the patent office. The government would dearly love to be able to complete this part of the process in eighteen or twenty-four months.

WHAT the HELL is a TM?

Trademark registration is a whole different process, although you should still use the services of a patent attorney. To that point, you will then be able to register a trademark based on your intention to use it in commerce.

All your attorney has to do is file an application along with a drawing of your trademark in its least specific form—i.e., block letters with no punctuation except for hyphens. If you plan some distinctive way of displaying it— some tricky typeface or with some sort of drawing built around it, for example—you have to furnish a sample of the name in that form, too. In the case of your going for an actual trademark registration and not an intent to register, you'll need proof that the product is currently being sold both in and out of state. (You needn't send that proof along with your application, by the way. You just need to have it on hand in case there's a question.)

Your attorney bundles up your application, along with a check for $325, and that's it. In a few weeks you'll receive a filing receipt from the Patent and Trademark Office, acknowledging that they got your application. Later, after a few months, you'll get a letter saying one of two things: (1) Your trademark seems to be allowable and will be published for opposition by anyone who thinks it conflicts with their trademark, or (2) your trademark is not allowable.

It might not be allowable because it's too close to an existing trademark. Or it might be in violation of some statute. Or it might be what the office calls "scandalous." (That covers not only obscenities but also things repugnant to public morality, such as the use of the deity in your trademark.) The patent office will cite its specific reasons for not allowing your trademark, and you can argue back if you disagree.

If, after all is said and done, no one objects to your trademark and the patent office can't find any reason not to allow it, you will be permitted to use the symbol "®" with your trademark, indicating that it is registered and, therefore, protected.

While the registration is still pending, you're entitled to use a "TM" with your trademark. This indicates to the world that it is your common law trademark and, by inference, that you plan to register it. It does not mean that the trademark is protected. That TM is a statement of *your* intention, not the government's.

Once your trademark is registered, no one else can use it and trade on your product's reputation by stealing its good name. The actual degree of protection is a little subjective, though. For instance, it might be possible for someone to use a name exactly like yours, provided that there is no likelihood that confusion will arise between the customers of goods bearing the same mark as to the source of the goods. It's all a matter of judgment, and if pushing comes to shoving, it has to be settled in court. Needless to say, the government won't register a trademark if there's one already registered in your class with which it would clearly be in conflict. So, by registering your trademark, you also get some assurance that you are not in danger of looking too much like some other trademark on the market. A trademark may be renewed every ten years for as long as it is used in interstate and intrastate commerce.

WHAT the HELL is a ©?

Copyrights used to confuse me. According to the old law, you had to publish something before you could copyright it. The question was, what constituted "publication"? If you wrote a poem and ran off some copies on the office copy machine, had you published it?

Worse, until you published, anyone could come along and steal your idea. (In theory, at least. In practice, the ancient "common law copyright" gave you protection of a sort by avowing that the creator of an idea owned the rights to it.)

The whole thing was a little murky for my taste. But as of January 1, 1978, a new copyright law went into effect in the United States, and it clarified a lot of that murkiness. Here's what the new law says:

As soon as you write your poem or paint your picture or do whatever it is you want to protect, the copyright comes into existence, and you can register it. You don't have to publish it first. As long as the material has been fixed in some tangible form, even if it's just a handwritten manuscript, you can get protection.

Better still, that protection now lasts for the lifetime of the creator, plus fifty years. (Under the old law, a copyright was good for twenty-eight years and could be renewed for another twenty-eight years. After that, it became part of the "public domain" and available for anyone to quote or reproduce.)

Copyright protection is the simplest of all to obtain. You simply file the official copyright application form with the Copyright Office in Washington. Not the Patent Office. The Copyright Office. There's nothing to it. And it costs only $30 for the copyright fee, plus a few dollars to the lawyer for filing it. No more than $100 on the average. In fact, you don't really have to have a lawyer in on the case, since it's no major undertaking to fill out the form yourself.

DON'T be a HAZY DREAMER

All through this whole chapter, I've been concentrating on the protection aspects of patents, copyrights, and trademark registration. But there are some equally important, if less obvious, benefits to filing your application early in the game. By applying for a patent, you're telling the world that you're serious about your invention. You mean business, and you're doing a full-bore, professional job of developing your idea and taking it to market. That can be very important to you. Millions of people have ideas, after all. Bright ideas, even. Most of them never bother to take that next step and apply for legal protection. Simply taking that step begins to separate you from the crowd of hazy dreamers and puts you in the much more select group of doers. Doers attract a great deal more attention—and money—than hazy dreamers.

Let's say, for example, you're approaching a manufacturer to whom you'd like to sell your idea for a royalty deal or a flat fee. If you have a patent in the works, you're way ahead of the inventor who doesn't. Your application for a patent is evidence that you've done your homework, that you're not just wasting the man's time with some half-baked scheme. It gives the manufacturer one more reason not to ignore you, and you stand that much better a chance of being considered seriously.

What's more, a reputable manufacturer will feel a good deal more comfortable talking to you about your invention if you have a patent application in the works. Strange as it may seem, that patent application protects *him*, too. Scott Kelley, my patent attorney, explained it to me this way: "With a patent application, you, as the inventor, are relying on the patent laws of the United States to establish your rights. Here is something that's on file in the patent office. It's a matter of record, and if you get into a fight in the courts, it can be established as to what you had at that time and what you submitted to them."

It makes sense when you stop to think about it. One of the things any manufacturer dreads is that an inventor may come to him with some idea, then claim later on that the manufacturer stole it. Manufacturers are always developing ideas, and they may well have something already in the works that related in some degree or other to your invention. But your patent application spells out completely, and in full detail, exactly what it was that you presented to the manufacturer for his consideration. It's all down in black and white with very little room for fudging. All you can prove about the discussion at your meeting is the invention you describe in your patent application.

By the same token, you've got proof of what it was you presented in case the manufacturer *should* try to steal your idea. The end result is a much more comfortable meeting. He'll feel more secure. You'll feel more secure. Together you can get down to the business of discussing your invention without all those doubts casting a shadow.

Finally, your patent application tells the manufacturer a couple of other things he likes to hear: that there's a pretty good reason to believe your invention really is unique and that you have some indication your idea can be protected. Those are two of the early questions you seek to answer in the process of applying for a patent. Your patent application tells the manufacturer that his investment in your idea will have some degree of protection.

The same kind of thinking applies when you make a pitch to get backing for your invention on your own. To a bank committee, perhaps. Maybe a group of investors. They're all impressed by the same kinds of things as the manufacturers. And for the same reasons. If they're going to sink money into your idea, they'd like to know that you're going about it professionally, and they'd like to believe that their investment has some protection. It's up to you to convince them.

I hope that the lady I met in the line reads this and follows up on it. I'd like to believe she will develop her idea, get protection, and wind up in Acapulco with a fortune. I hope I've done a good enough job of convincing her—and you—that getting legal protection is one of the earliest things to think about when you're developing an idea. But I can't leave you without pointing out an important fact. A patent, copyright, or trademark registration is absolutely your only protection. But the protection is *not* absolute. There is a dark side to the coin, and I'll devote the next chapter to it.

Research
AND
Rip-Offs

*Those "Inventors Wanted"
ads may be selling real help or
be just a scam. Here's how to
tell the difference.*

IN THE OLDEN DAYS, ROBBERS STOOD BY THE ROADSIDE RATTLING
their pistols and waving their sabers, yelling, "Stand and deliver!" These
days you'll find them back among the aspirin ads in your newspaper, saying,
"Inventors Wanted."

Most inventors are wary enough of the knock-offs. But many new inventors don't realize that there are other kinds of scoundrels out there who may
be even more dangerous. These varmints don't even bother to run off with
your idea. That's too much trouble. They go straight for the wallet.

In case you haven't run across this particular scam yet, let me fill you in
on a couple of variations of the "Inventors Wanted" rip-off.

TRUST ME . . .

The first variation is the "don't worry about a thing" gambit. You submit
your idea, and if it meets their standards for a marketable invention (I've
never yet heard of one that didn't). Your idea *always* meets their standards, no
matter what the invention is), they'll "market" it for you. The unspoken

promise its that they'll do all the dirty work for you and lead you through the business jungle to vast wealth and eternal bliss in the invention business.

What really happens is that the rip-off company copies a sheet of the latest batch of inventions they've accumulated and sends it out to a few manufacturers and marketers. In the first place, there's only the briefest kind of description of the product. In the second place, I wonder who gets the fliers, since I've rarely met anyone in the idea business who has seen one. And in the third place, you can imagine the kind of reception that those sheets get when they arrive at some company. If you can't, let me give you a few quotes from people I talked to in the process of researching this book. They're instructive.

A Sears, Roebuck and Co. buyer: "I have never in fifteen years in housewares and shoes even *heard* of a manufacturer who has been contacted by one of these businesses, and I've dealt with hundreds of manufacturers over the years."

A vice president, creative services, Hallmark Cards: "I've never been approached by a new-product-development company."

My own question is, who *has* been contacted by these rip-off artists, if two of the biggest idea users in the world have been skipped? But the question is academic, based on the response of some people who have gotten the fliers:

A vice president, new products, Fisher-Price Toys: "In maybe twenty-five years at this company, we have not been approached on a professional level by any 'Inventors Wanted' organization. They send form letters from time to time, but I throw them in the wastepaper basket."

I've never yet run across a single instance of an "Inventors Wanted" company *ever* making a sale of an idea to some manufacturer or marketer. To be fair, I'm sure there are some. I've never yet seen one of them even try hard. The most concerted effort at selling I've ever seen them try is that tired old flier. Today, they of course offer up a Web site as well. There may be some reputable company out there doing this kind of thing, but I'd sure like to see one. In fact, I'd like to hear from one so I could recommend them when the question comes up. Incidentally, 149,000 copies of my first version of this book were sold. In all the years it was in print, not one "Inventors Wanted" marketing company accepted my invitation. What does that tell you?

That's one version of the old rip-off game. It's fairly transparent, although it may get past the inventor's defenses by making him feel really wanted. (These companies always profess to be impressed by the invention, firmly convinced they'll make a sale, ready to go to the mat for *this* idea—sign here, please.) Oh yes, there's *always* a fee! Usually a few hundred dollars. To start.

BEWARE. BEWARE. BEWARE.

Another variation of this scam is a little less transparent. In fact, until you take a close look, it even seems like a reasonably decent deal. Here's how it works:

A "marketing" company looks at your idea, gets excited, and offers you a marketing survey and analysis of your idea. It sounds like good stuff. They promise to give you a complete breakdown of the market for your idea, full background information on your product category, professional drawings of your product, cost analysis—all the things you'll need to make a sale to some manufacturer.

That's the theory. The practice is a little less impressive. At best, you'll pay a stiff fee (up to $8,500) to get a handsomely bound document that purports to be a complete analysis of your product and its prospects in the marketplace. The document will be fat and impressive, chock full of genuine numbers and, in several places, will actually mention your product by name. The drawings will be bound into the volume so that they fold out dramatically. Hot stuff! But what does it all mean?

What it usually means is that if you sign up, you've been ripped off. Most of the information is totally irrelevant to your product. Anyway, you could find it with a quick trip to the library. There may be page after page of nonessential information lifted directly from the last census report and having nothing whatever to do with your product. An analysis of your product category may be nothing more than a long-winded rehash of someone else's research on the number of stores selling that kind of product.

Worst of all, these reports invariably pump up the inventor's ego by telling him what a swell idea he has and by predicting success (in carefully circuitous language) no matter what the product is or what the real odds of success. On the basis of this impressive, expensive, and meaningless report, an inventor may be moved to throw away thousands of dollars and months of effort in pursuit of a phantom. The idea may not have a whisper of a chance on the market, but in guarded language, this will lead the inventor to believe he has a real winner. It's criminal! Or it ought to be.

When I was planning this book, I had a fantasy of sending my worst idea to one of these companies and publishing the results of its report, just to point out how absurd this rip-off is. As it turned out, I didn't have to. A couple of guys I know did it for me. To keep me out of court and them from possibly being harassed, let's call them Fig and Newton. Fig and Newton worked for months developing an electronic game we'll call Spots. Now, new games are *poison* in the new-idea marketplace. Toy and game manufacturers have so many new ones going all the time that they don't need any outsiders' ideas. And any outsider who tries to beat them in the marketplace on his own is likely to get cut into little tiny pieces (and then diced again) by the intense competition. Trying to market a new game on your own is the nearest thing I can imagine to financial suicide.

Worse still for Fig and Newton, this particular game would have had to go at retail for more than $60—a kiss of death for a game. The two inventors, being new to the business of inventions, had no idea of all this, nor any way of knowing. They just went about their business, developing their new game. (As a matter of fact, it's a pretty good game, too.) Then they answered one of those ads. Sure enough, for a fee of nearly $5,000, they could get "the complete market analysis" of their game. They wrote the check and, in due time, received the report. It's a lulu.

It's handsomely bound in a hard cover with the marketing company's name embossed on the cover. Very uptown looking, until you look closer and realize it was done on one of those desktop office bookbinding machines. It has 149 pages—including several blank filler pages. It has neatly tabbed sections with headings like "Product Definition" and "Market Evaluation" plus five appendixes. And it has numbers, numbers, numbers. From time to time it even says "Spots" in the copy.

As you may know, there are machines that will insert any given name from time to time as it goes through a predetermined piece of copy. You've probably received letters typed that way from land developers, dance studios, and contest promoters that say, "You may already be a winner, Mr. Smith." That's how the report was done. Somebody put together a bunch of standard sections, plugged in the product name in a few places, did some drawings and a dab of real research, bound it up in a hard cover and shipped it off to Fig and Newton.

To show how bad it really was, let's be specific. Of the 149 pieces of paper between the two covers, 133 are meaningless in relation to the actual product under consideration. There are eleven pages devoted expressly to Spots, including three pages of drawings that are professional but also very vague. You could have had them done for about $100 apiece.

In addition, there are five pages on which the name "Spots" is inserted into totally generalized material. I think it would be safe to conclude that the really pertinent part of the report could have been printed on not more than sixteen pages. If it weren't double spaced (and the entire report was double spaced, with generous white areas between sections and paragraphs) the report could have been done on less than eight pages. My first conclusion has to be that this is a lot of money to spend for eight pages of information.

For instance, there are three pages devoted to an analysis of the market for outdoor play equipment. Outdoor play equipment! What has *that* got to do with an indoor electronic game? It looks to me as if the clerk who pulled together the prewritten sections to make up the report simply went down a shelf pulling out every section that could remotely be connected with games. A report on any other toy or game would have contained substantially the identical material. And it would have been as consistently bad!

Another example of a filler in the report is the six pages of blank paper stuck into one of the appendices, just to pad it out a bit and make it seem impressive. On closer examination of the whole report, though, I'm inclined to believe that these blank pages do less harm than some of the ones that have words on them.

Where the information pertained to the product at all, most of it was extremely misleading. For instance, there's a fairly workmanlike breakdown of manufacturing costs. But the tooling cost was spread out over 200,000 units of the game. The poor inventors, looking at that bit of information, can only conclude that they were working on a game with a selling potential of at least 200,000 units or more than $12 million worth at retail! Holy cow, Fig—we're gonna sell $12 million worth! Now that's possible. Anything's possible. But it's not the kind of assumption you would want to bet your savings on.

And there's another, worse part to the report. By its *apparent* heft and completeness, Fig and Newton were convinced they had a valuable sales tool to take to game manufacturers along with their model and plans. Had it been a professional analysis they would have been right. Since it wasn't, they were wrong. If they tried to take that report to some manufacturer to make a royalty deal, they'd probably never get past the receptionist.

At one place in the report, the development company made the following marvelous statement: "Games have a healthy future predicted for them by industry representatives. Inevitably, game prices will increase due to rising costs of materials and an estimated 10 percent boost in operational costs for game manufacturers. But the increase in business should counter this."

Are they saying that there will be a loss on every unit, but it will be made up in volume? What, exactly *are* they saying?

Elsewhere there is a statement that "Spots seems to have appeal for this market and *should do well,* once manufactured and distributed." (The italics are mine.) Well, you'd better believe this is what any inventor wants to hear. No one wants to be told that his children are ugly! Especially when one's paying so much for an opinion.

You and I know—and so do the two inventors of Spots now—that the odds against success by any game are fantastic. But here's the company blithely giving the impression that this game is practically a surefire winner. (Notice how carefully they avoid making any concrete promises, by the way.)

There's one more quote in the report that's just too good to pass up. In the "conclusions" section of the book, the rip-off company says, "This item seems to be well conceived and of good design."

Well now, in the first place, "this item" doesn't even deserve the printer treatment of plugging in the name "Spots." It's so far into the report that the compiler probably figured the reader would already be asleep.

Next is "*seems* to be well conceived." No chance of being sued for that ambiguous statement. Ben Franklin built a solid reputation on his use of the word "perhaps."

For the money this report cost, I would have demanded some positive (or negative) statements about my product based on the best research available at the time of its preparation. Instead, what the inventors got was a lot of warmed-over census reports and some generalizing about the marketplace. No discussion of the odds against such an idea, no guidance as to what steps to take next, nothing of any substance at all. After reading through the whole report, I can only come to the following conclusion:

The product development and marketing services company that prepared the report didn't care about anything other than making a profit on the time spent with the inventors. That they did, handsomely. Just for kicks, I analyzed the report and tried to come up with an estimate of what costs actually were involved in producing the $5,000 report. Here's where I came out:

1. One hard cover with imprint (ordered in
   Quantities of 1,000 I assume)....................................$ 5.00

2. Fifteen different plastic tabbed divider
   Sheets (ordered in quantities of 1,000) ....$ 3.00

3. Collation of 149 individual pages
   For one report ....$ 8.00

4. Typing eight pages of special copy ....$ 12.00

5. Three professional drawings.....................................$ 300.00

6. Typing five pages of insert copy ....$ 16.00

7. Conference time for total report including
   writing and signing a one-page personal
   cover letter. (Six hours @ $200 per hour) ....$ 1,200.00

8. Printing six copies each of sixteen pages
   with specific product references ....$ 20.00

9. Twenty percent for overhead and coffee...................$ 312.80

   TOTAL...$ 1,876.80

As you can see, the company made more than a nice profit . . . for themselves. But that's not really what bothers me. Rather, I'm concerned about the misleading information. Such companies ought to be held responsible for the inaccuracies of their conclusions (even though they were carefully phrased to avoid making any commitment). Inventors who are encouraged by such reckless suggestions can wind up wasting untold amounts of sweat, toil, tears, and treasure chasing a will-o'-the-wisp they'll never capture. Or recapture.

I'm not trying to talk anyone out of marketing his idea, believe me. Not even a toy or electronic game idea, although you know by now how I feel about the odds against success in that area. My only concern here is that inventors who fall into the clutches of these rip-off artists may never be apprised of the odds against them until they find out the hard way. In the example we just saw, it is a gross injustice to the two inventors to tell them that "Spots seems to have appeal for this market and should do well."

If you're tempted to go with some company that offers to market your invention, be careful. One way to be sure of their intent is to insist that they take their fee as a percentage of what *you* make. It may cool their enthusiasm considerably to be told they won't make any money until you make money.

My advice is to spend the same amount of money on a sample production run or a patent attorney or some valid research or something else that will help your product along in a realistic way. Inventions never have enough backing anyway. So why waste what finances you have on some outfit that claims it will take the drudgery out of being a successful inventor? I know a lot of successful inventors—I'm one myself—and I've never yet heard of an easy way to go about it. Drudgery goes with the territory, and no rip-off artist is going to be able to change that, no matter what he or she claims.

Since I've just destroyed a $5,000 report on the grounds that the information was shabby, I thought it only fair to show you an example of a $10,000 report I feel was worth the price.

THE RIVERBOAT GAMBLERS

Several months ago, I was introduced to a couple of guys who (you guessed it) had a "great, new, million-dollar idea." They wondered if I could help them market their product. Unfortunately, the product was an automotive sound system—something about which I know less than the reproductive progression of a Crossopterygian.

Having disqualified myself as any kind of authority on their subject, I then proceeded to tell them something I've already told you: Apply for a patent. They had. Good.

"Now, go license the idea," I suggested. "Don't try to make and sell the product on your own."

"Why?" they asked.

"Because," I guessed, "you're looking at extensive tooling costs—somewhere around $40,000 for openers."

"Actually, we checked that all out, and the figure is $32,500," they shot back.

I obviously didn't have a Fig or a Newton on my hands this time. That figure didn't deter them, so I laid on additional cost estimates. I pointed out our fee for naming the product, designing the packaging, creating the necessary marketing support tools and helping establish a distribution network would be around $70,000. They had figured around $65,000. Nothing spooked these guys. Real Riverboat Gamblers.

"The miscellaneous stuff will probably run your initial investment up to $120,000," I added. "And at that you will have no assurances of success. Worse yet, if you're successful, some giants in the electronics world will jump on your concept and kill you in the marketplace with a variation on your theme."

The $120,000 they had thought of. Being knocked off was a new concept. After we discussed their being ripped-off in detail, they started to reconsider their position. At that point they left.

Then they came back. They had decided to become manufacturers after all, and wanted to know how soon we could come up with a catchy name. At this point, Don-the-wet-blanket stepped in again with, "You're not ready to spend money with me, yet. How much have you budgeted for market research?"

"None." They wondered if they needed it. I pointed out that if I were contemplating a $120,000 initial investment in a brand new concept, I would certainly want at least a few more bits of information to help me decide to GO or NO-GO.

It pained me that I was sitting there doing my best to boot away a $70,000 marketing project, but I gave them my lecture on research. First, I said, there is no better research in the world than putting a prototype run of a product on a store shelf and watching what happens to it. Anything south of that is educated guessing. When the cost of good research is about the same as doing a prototype run and actually selling the product, I'll go for the sales experience every time.

In the case of the Riverboat Gamblers and their automotive sound system, however, the cost of going into a limited market was much higher than the cost of research. I figured—correctly, as it turned out—that they could get a good preliminary market research report for around $10,000. I suggested that they go to a reputable marketing research firm and commission them to do a market study. These firms do specialize in research and are not offering 101 other services like the marketing submission firms.

These firms will have good references, i.e., clients who have used them; and they will belong to organizations in their industry; and they will be in good standing in these organizations.

Check them out, and they will tell you the methodology they're going to be using in their market research study.

The Riverboat Gamblers took my advice. What they got was worth every penny they spent. There were no absolutes in the market study research. There never are, not for $10,000, nor for $50,000,000. But there was some sound, fundamental information about the market for their product and how best to approach it. The very first words of the report were heartwarming to me. "Purchase interest for the product is quite low based on the concept description of it."

No phony hype about what a wonderful product you have there, Mr. Inventor. No glowing and unsupported prediction for sale success. Just hard facts—some encouraging ones and some discouraging ones.

The researchers didn't fill up their study with a lot of canned statistics. Instead, they took a handmade sample of the sound system out to some shopping centers and talked to potential customers about it. They demonstrated the sound system and probed reactions to it both before and after hearing the demonstration. And the questions they asked were meaningful.

As the researchers said in their report, they had a specific set of objectives in mind when they questioned their market sample. "The purpose of this investigation," the report said, "was to determine the market viability of the product, to identify its strengths and weaknesses in the eyes of potential customers, and to identify the target market. Specifically, the objectives of the study were:

1. To define the prime prospect in terms of its demography—age, income, type of vehicle owned, education, type of sound equipment owned in auto and home.

2. To determine what is liked and disliked about the sound system.

3. To determine how the sound system compares to alternative systems.

4. To determine at what price individuals feel the system should sell.

5. To obtain any purchase interest in the system.

6. To determine at which type of retail outlet consumers would expect to purchase the system.

7. To rate the system on a variety of attributes:
   —sound reproduction compared to other speakers
   —design attractiveness
   —ease of installation
   —sound separation

9. To obtain consumers' perceptions of products to which the new sound system is most analogous."

It's apparent that the researchers were out to give the inventor of the system some basic knowledge about who to sell to, how to sell, where to sell, at what price to sell, and even how to improve the product's marketability. The researchers may have laid it out in dry-as-dust prose, but it's poetry to the ears of anyone who's about to sink $120,000 into a gamble on his new idea.

The guts of the report take up about 30 pages and contain more useful information than a dozen 149-page rip-offs like the one Fig and Newton bought. Just to give you an idea of the difference between the reports, here are a few examples of what the good one had to say:

After pointing out that the test subjects expressed much more interest in the product after hearing it demonstrated, the report went one step further and probed the importance of that fact in terms of retail price. "The expected price for the product increases substantially after listening to it . . . The expected price is about $99 before listening and $129 after listening. Store owners tend to hold to the expected price before and after listening."

That gives my friends, the Riverboat Gamblers, a pretty fair notion of how to position their product in the marketplace and how to sell it most effectively.

The researchers chose only people who told them they would like to improve their automotive sound system—in other words, the most likely customers for the product. Then they analyzed just who it was who expressed that interest. "The individuals who qualified for this study . . . are definitely younger, more likely to be single, to be students, and to have lower personal incomes. Stereo owners are more likely to exhibit these characteristics than monaural owners . . . Although [younger singles and students] have low personal incomes, the price of the unit is not a problem because they have a great deal of discretionary income they can spend on things that are personally gratifying and enjoyable to them. Lifestyle is more important than income in defining the target market."

With that piece of information in hand, the Riverboat Gamblers know who they'll have to reach with their advertising and promotion for the product.

Another important observation about the good research is that the researchers laid all their cards on the table for their clients. In a section titled "Methodology," the researchers explained where they set up their interview project, who they talked to, how many people they interviewed, what questions they asked, how they went about it all, and how they analyzed the data. Needless to say, the people who did Fig and Newton's "research" gave out no such information.

ON CONNING the PROS

There's one last thought about research to consider. Since this chapter is about a pro and a con, let's see how the right kind of research can be used to con the pros. We'll presume for a moment that both Fig/Newton and Riverboat Gamblers, Inc. have decided to try selling their ideas to a manufacturer. Fig and Newton's $5,000 investment in research will be laughed right out of any boardroom because it's bad, slippery stuff. The Riverboat Gamblers, on the other hand, will have a valuable supplementary tool that will help convince a manufacturer that the idea they have is worth further exploration. When push gets to shove, the Gamblers will be able to tack an additional $10,000 onto their deal to cover the cost of their research because it's good stuff that the manufacturer can use. The Riverboat Gamblers will be repaid for spending money to help make a sale. Good for them. Bad for me. Undoubtedly, the manufacturer will have his own marketing group that will be eager to use the $70,000 for naming, packaging, and the like.

But that's my problem, not yours. Your problem is how to tell good research from the rip-offs. I hope that by now you've developed a healthy skepticism about those people who promise you success and deliver a census report.

Doing
YOUR
Homework

Calculating the wholesale and retail price, including manufacturing, packaging, selling, and shipping

SOMEWHERE ALONG THE LINE, I SEEM TO RECALL PROMISING YOU that you were going to have to work like a dog and spend real money before you turned your idea into a success. Well now's the time to start.

The little bit of effort and money we talked about in the chapter on patents was just the beginning. You see, there are still many unknown quantities about your idea at this point. It doesn't matter whether you plan to sell the idea to some manufacturer for a royalty or make or sell your invention on your own. One way or another you'll have to turn all those unknowns into knowns before your product reaches the shelves of the stores.

You'll have to find out if your idea actually can be manufactured. You'll have to figure out how much it will cost—including materials, manufacturing, packaging, shipping, sales commissions, overhead and, of course, profit. You'll have to get some kind of handle on whether or not it will sell. And how well. And for what price. In short, you'll have to do some running, digging and spending.

Let's tackle those questions one at a time, please. I confuse easily.

THE REAL THING

Can it be made? The answer to that, logically enough, is to make it. A sample production run, perhaps. Or some prototypes. Maybe just a model. It depends mostly on the kind of idea you have. If it calls for tooling up with thousands of dollars worth of special dies or something, it's a pretty good bet you're not going to be making any sample production runs. But in that case you can make a model of your invention—or have someone make it for you. In the case of Rickie Tickie Stickies, the idea was simple enough and the product was inexpensive enough that running off a sample batch of 3,000 was no big thing.

In between the two extremes, there's the possibility of making a prototype. That's a real, working example of your product in the same shape, size, and detail as the production version will be. If possible, it ought to be made of the same materials, too.

Until you do one of the three things—a production run, model, or a prototype—your idea is a pig in a poke. It's a figment of someone's imagination. Your imagination. Nobody has proved it will work. Nobody knows if it actually can be manufactured.

But once you have a sample, you don't have to wave your arms and say, "Imagine, if you will . . ." to explain your invention. You can just hold it up for people to look at and touch. It exists in living, three-dimensional, full-color splendor. It's the real thing.

Now, that may strike you as a lot more trouble than it's worth. But it's an important phase in the life of any new product. It comes under the general heading of getting that pig out of that poke. It begins to solve some of the problems of manufacture. No matter what your idea is, little details you wouldn't have anticipated in a million years on paper will leap out at you when you have to translate your idea into a three-dimensional object.

Sooner or later, these snags are going to have to be dealt with. Much better you should say "Oooooops!" now than when you may be up to your fiduciary nerve in production deadlines. If you're going to manufacture and sell the products yourself, it's just one more step bringing you closer to success. If you're planning to sell your idea to a manufacturer, he'll be very happy to have those problems already solved. It shortens the time he has to spend between buying your idea and getting it on the market. A prototype also helps a lot in your patent application process. Remember, it's in the marketplace that your idea will generate money, not in the production planner's office.

TAKE my PRODUCT, PLEASE

Another critical question has to be answered now. In fact, it's really a fiendishly complex, interlocking series of questions. The series begins innocently enough. How much will your invention cost to manufacture? That question quickly leads to the question of how much your invention will have to sell for at retail. That, of course, leads directly to the ultimate question: Will anybody want to buy it?

Let's take a look at all the costs involved in manufacturing your invention. You *have* to find out what it will cost before you talk intelligently about your invention. Many beginning inventors regularly ignore a number of real costs they'll have to face, and it's a trap. By doing so, they con themselves into thinking they've got a much more attractive proposition in their hand than they really have. The costs to calculate are these:

THE COST of MATERIALS

You can't just say—as one inventor recently told me—that this is negligible. "Oh," she said, "there's nothing to it at all. Just a piece of wood and some paint and stuff. Less than a penny apiece." But when you multiply "less than a penny" by a hundred thousand or so, it becomes a substantial cost. Calculate it as closely as you can. Think in terms of mills . . . tenths of a cent . . . as the pros do. Call up or go see some suppliers and find out from them what they'll charge for a supply of the materials. Get quotes from several suppliers. And get them to quote on the quantity you'll need for a small production run, then for a larger one, finally for the biggest production run you could realistically expect. Make sure they understand exactly what grade of material you're looking for. (You may not know, yourself, at this point. But the suppliers can help you figure it out, often as not. After all, they're looking to make a sale, so they have a vested interest in being helpful.)

PRODUCTION COST

I'm talking about the physical activity of making your goods. If you're going to hire an assembler to do it for you in his shop or factory, you'll have to go around getting bids—the same as you did when you were pricing materials. If you're planning to make the goods yourself, you'll have to calculate the cost of the space you'll use as your factory, the cost of the equipment you'll have to buy, and the cost of labor in the manufacturing process.

By now, you may be wondering how you—probably not the world's foremost model maker and certainly no manufacturing expert—are going to do all the things I'm talking about: make a prototype or a model of your invention, calculate productions costs, figure out what materials you'll be using, and so forth. The answer is easy. Don't do it yourself; hire an expert. Get out your Yellow Pages and look up "Industrial Engineers" or "Industrial Designers." These people make their living by working out the proper materials, the best processes, and the least expensive techniques for manufacturing things. They can also make your model or put you in touch with someone who can do it. These people know all the ins and outs of manufacturing and can find ways to cut your costs substantially, so the money you spend on their services will probably pay big dividends. (You may be able to retain their services without putting out any cash, by the way. More about that in a later chapter.)

While you're wrestling with production costs, there's a nasty word you'll have to learn. It's "Amortization." It's a word that turns a solid, upright, hard-working businessman into a hollow-eyed crapshooter. Amortization means laying off the cost of your capital investment (your manufacturing equipment and factory investment, in this case) against the number of products you'll make with that investment. In other words, you have to guess how many units of your invention you're going to sell. Of course you can't. That's why it's a gamble.

The best you can do is make an estimate based on how many stores you think you'll be selling in, how many products like yours they sell every year, and how long you think you'll be manufacturing the product. Based on all that, you close your eyes, grit your teeth, and take a stab at some logical-sounding number. Be careful! If you try to write off the whole investment against the first few orders for your product, you'll jack your per-unit cost up so high that you'll price yourself out of business. On the other hand, you can't just airily figure on making a zillion pieces, either. You may never make that many units, and you'll be stuck with an expense that was never truly reflected in your cost of goods.

PACKAGING

Hardly anything is sold without packaging, so somebody's going to have to design a package for your product. That's going to cost some money. How much money depends in part on what kind of package you choose—a blister pack to hang on a J-hook rack, a box, a bag, or whatever. Unless you are an accomplished writer/art director/designer, you'd better figure on paying somebody to do it for you. As a starting point for your calculations, you can use $5,000. That should bring you right up to the point of being ready to print the boxes. It ought to cover artwork, layout, copy, typesetting, and mechanicals. You can get it done for less. You can spend a lot more. Five thousand is just a nice, round, rough starting point. Later we'll also discuss ways of getting this kind of work done without putting out hard, cold cash. But the expense still has to be reflected in your cost of goods, one way or another. In other words, you have to—excuse the expression—amortize it.

Once you have the design, you have to get some estimates from printers on how much it will cost to print the packages. There's no way an amateur can calculate this cost on his own. It involves mystical subjects like color processes, stock weights, coated vs. uncoated stock, and a hundred other things you don't even *want* to understand unless you're a printer.

Finally, somebody has to take the product off the assembly line and stick it into a package. It was at this stage of the game that we made a marvelous discovery with Rickie Tickie Stickies. We were dealing with vinyl decals, you'll recall, approximately nine inches in diameter, which were being shipped to us in cartons of about 3,000. That's a lot of slippery little pieces to handle. With some friends and neighborhood teenagers, we packaged the first batch in their plastic bags, working around folding tables in the garage. But very quickly we were swamped. Where to turn? Where else! We got out the Yellow Pages again and looked up companies that do packaging.

What looked like a monstrous job to us was really too small to interest the professionals. Oh, they'd do it all right. But in order to make our little job worth their while, they'd have to charge a lot more than normal per package. Then we had a brainstorm. We knew that certain organizations around the country hired physically challenged or mentally challenged people for packaging work that was not too demanding. So we called around and we ended up working with Goodwill industries in Long Beach, California, an easy thirty-minute drive from our house.

They'd handle a small job at rates competitive to what the professionals would charge on the basis of a big job. For several years it was a perfect blend of product and labor force for us. What was just as important, the relationship provided us with many touching, funny, wonderful experiences. All the kids who worked for Ruby Goakes, manager of the shop, had a fierce determination to do a successful job for us. No overload seemed to bother them, nothing got in the way of doing a first-class job for us.

For one thing, of course, Ruby and her kids were happy to have the business. Most of their work came from machine shops. Having a chance to work on something clean and bright like our flowers was a treat for them. Depending on their ability, some of the workers could only count out three flowers of the same color. Others could count up to nine assorted colors. Some learned to work with the heat-sealing equipment we brought into the shop. All of them loved their work and were pleased to be able to take home an occasional bright "reject" from their day's work.

The point is, you can do yourself and the world around you a very good turn if your product can be packaged, or even manufactured, by one of these "sheltered" workshops like Goodwill Industries. You'll be pleased with the quality of work, you'll get a competitive price, and some deserving people will get a chance to participate in society. That's a pretty fair parlay!

The cost of this packaging operation must be included in your cost of manufacture, naturally. And while we're on the subject of packages, don't forget that you have to ship your products. That probably means buying corrugated cardboard shipping cartons, each big enough to hold a quantity of your exquisitely designed, beautifully printed packages and strong enough to protect them from the ravages of the shippers. These boxes come in standard sizes, or you can have them especially made for your product. Here again, figure out exactly what you need, then get bids from several suppliers, based on different levels of production. After you've found your cost per box, divide that price by the number of units you'll ship in each box, and you'll arrive at your cost per unit.

The total of all these fixed costs is called your cost of goods or COG. But that doesn't mean you're done yet. It simple means you're done with the easy part. Now you have to get into an area of sliding costs, costs based on a percentage of your selling price. First I'll explain the costs. Then we'll figure out how to go about calculating them.

SALES COSTS

Somebody has to sell your products for you, unless you're going to spend the rest of your business life working out of the back of a station wagon. These salesmen are paid a commission, or a percentage of the wholesale price they get.

Later on we'll get into the whole question of reps and commissions. For right now, just figure on paying 20 percent of the price of your products to the salesmen. (It may be less than that, in real life, depending on the type of distribution you choose. But twenty percent is a good safe starting point.) That commission is based on the price *you* get, not the price the retailer gets when he sells your product in his store.

G&A

That's technical talk. It stands for "general and administrative" costs, but most businesspeople like you and me both refer to it by the initials. It's also called "overhead." Or both. It includes a mixed bag of costs, things like salaries, advertising and marketing, your own office overhead (as opposed to the manufacturing overhead), all of the little niggling, miscellaneous costs of being in business. Like insurance for fire. Theft. Liability. Earthquake. Hindenburg. How much is the G&A cost? The only way to calculate it exactly is by amortization. But as a general rule, you should allow no less than 20 percent of the wholesale price of your product as G&A expense.

That may sound astronomically high when you're working out your cost of goods right now, but later on when some of the bills start rolling in, you'll understand that it's pretty realistic. Even if you're planning to start on a shoestring and work out of your home, allow for this expense. Sooner or later, if you're successful, you'll be expanding into a real office, just like the big kids. When you do, you won't have to jack up the price of your product to pay for your new prosperity if you've already included it in your costs.

SHIPPING

Shipping cost is determined by how big the box, how heavy the product, and how far the journey. Normally, you expect your customer to pay for the shipping, but for a number of reasons you may find that you'll wind up paying at least a portion of the cost yourself. Therefore, you ought to know what it is.

When we first got involved with shipping, it didn't take us long to realize how much we take for granted the fact that parcels leave one place and arrive, magically, at yet another. Once we passed the stage of handy home delivery, however, we suddenly found we were in another new arena of which we were supremely ignorant.

We first learned that you have to be careful about the carton you use for shipping, because practically every company has some restrictions on what it will handle. I remember having a length of chain in the shipping room that established the allowed measurement of girth-plus-length of our orders. We also found that we could buy a shipping scale that had destinations coded by distance and made computing of charges much easier.

While we were learning about shipping—under the tutelage of all the companies that were after our shipping business—we discovered something of the psychology of shopkeepers. Shopkeepers, it turns out, never place an order for your product until the last one in inventory has been sold. Then they leap to the telephone in a blue panic and beg you to ship immediately because they have a line at the counter waiting for your product to arrive. That means you're going to have to learn all the fastest ways of getting an order out, as well as the cheapest. (We also found that shopkeepers never change the size of their order. If they bought a dozen the first time, they'll always order a dozen. Even if they're reordering every twenty minutes, it never seems to occur to them to place a bigger order. Why? Damned if I know!)

This led us to the discovery of air freight. At first, we worried that the cost would be prohibitive. But it was clear that we were losing quite a few sales by not having our product on the shelves of stores that ran out. So we did what had become our custom. We called air-freight companies and got the story on what we would have to spend for their services.

You'll probably find, as we did, that you'll use a combination of shipping processes. We finally settled on a combination of UPS air service, regular UPS, general air freight, and sometimes surface trucking for practically all of our shipping. We avoided the U.S. Post Office whenever possible. A few experiences convinced us that each post-office branch keeps a truck on hand specifically for the purpose of running over all the boxes at least once on departure and again on arrival. This did not please our customers at all, and they usually let us know.

I mentioned that you might find yourself paying at least part of the shipping cost yourself. One of the most important reasons for doing that is to use a shipping allowance as a sales incentive. You offer to pay part of the cost for a customer if he'll place an order for some minimum quantity of your product—more than he would have ordered without the incentive. Customers with well-oiled financial management like that because they can turn it into a real savings for themselves. You'll like it because it leads to bigger orders. There's no real set figure, but as a rule, you allow 3 percent of the price of an order as a shipping allowance.

There are some customers who jam freight allowances down your throat. These are some of the big, corporate-style customers who write their own freight contracts and you like it or lump it. It can lead to some soul searching on your part to see if that huge, exciting order from that huge, exciting chain of stores is really worth it. By the time some of these customers automatically refuse to pay freight, automatically take a 2 percent discount for each cash payment even though they don't pay for sixty or ninety days, and automatically return unsold goods for credit even though you never agreed to it, their orders become a little less exciting and a little less-worthwhile.

No, they don't all behave that way. But you ought to be a little cautious and do some checking when the demands start going beyond the bounds of common decency.

While shipping is a headache at times, it is one of the most important aspects of a manufacturing company. Someone in your organization had better make it his business to learn all the ins and outs. The freight business is a highly regulated one with rigid policies. You have to keep informed not only about rates but also about the whole question of liabilities and damages. You also may find that you have to educate some of your customers, too.

Once we had a particularly knotty bit of education to do. We had worked out a system whereby we air freighted a bulk shipment to each of several cities. Once it was unloaded from the plane, the shipment was broken down into individual orders and delivered to our customers by UPS. The problem arose from the fact that our invoice made separate note of the shipping charges for the order.

That was all well and good, except that UPS put its own stamps on each order for the amount it charged to deliver from the airport to customer. There was nothing on the package to denote the additional cost of the air shipment to the point where UPS took over. So a customer would get an invoice charging him $3.00 for shipping and a package that said $1.62. We got some pretty heated phone calls about how we were gouging people on shipping. It finally got so bad that UPS had to print up a batch of fliers explaining the situation. One of the fliers was delivered with each order and a while later the howls of protest finally subsided. But it was pretty ugly for a while.

PROFIT

Ah yes, profit. Profit is like one of those optical illusions that keeps changing perspective while you look at it. On one hand, it seems a pretty puny return for this idea of yours. (That's when you're looking at the few pennies you'll get for every product you sell.) On the other hand, when you're trying to shave your cost of goods to arrive at a competitive price, it seems like an inordinately huge chunk of cost. You're right on both counts. Profit is simultaneously way too high and nowhere high enough. For whatever it's worth, most companies are happy to make between 10 and 15 percent of the factory selling price of their product as profit, before taxes.

Don't get the idea that you're limited to that, though. Some products can be "value priced." That is, the apparent, perceived value of the product is very high in relation to the cost of making it. For years, kids' toys with endorsements by the reigning cowboy star fit into that definition. You can plug in any number you want as profit, so long as the retail price still makes sense to the consumer. If it works out that you're making 3,000 percent profit, that's your good luck—as long as it sells.

ANOTHER ABSOLUTE PERHAPS

And that's it. Those are all the costs that go into making your wholesale price. As I mentioned, some of those costs have to be arrived at backwards. You've got some hard, fixed costs: materials, production, packaging, and so on. Then you have some costs that are figured as a percentage on your selling price. But how can you do that if those are the figures you have to use in calculating that price? Sounds like Catch 22. Since this is starting to get confusing, let's see if an example can help clarify this whole process.

We'll start with a hypothetical retail price. That's the price the ultimate consumer will pay for the product. How you zero in on that figure is covered a bit later in this chapter. For now let's say the retail price is $2.00. The first thing we do is subtract the markup for the retail store. On average, that amounts to about 50 percent of the retail price, or $1.00 in this example. So now you're down to $1.00. That's your wholesale price. From that we subtract the sales or distribution and shipping costs which average about 20 percent or $.20. Your $2.00 retail item brings your company about $.80.

From that you subtract your cost of goods. Let's say your product costs you $.40 to make. We now have $.40 left to split between G&A and profit. Using our twenty percent of wholesale for G&A we come up with $.20. That leaves you $.20 before-tax profit, which on a $2.00 retail item is about right. Here's how that should look on your notes when you're doing your first financial review:

$2.00	Retail
−$1.00	100 Percent Markup
$1.00	Wholesale Price
− .20	Twenty Percent to Sales and/or Distribution and Shipping
$ .80	Factory Selling Price
− .40	COG
$ .40	
− .20	Twenty Percent of Wholesale Allotted to G&A
$ .20	Before-Tax Profit

It's that simple to do when you're using a hypothetical example. In real life, there's a little more to it. For one thing, you have to do some refining of the cost figures you'll plug into your calculations. So don't break out the scratch pad just yet. You've got some legwork to do.

DISTRIBUTION

What you want to find out now is the best way to distribute your product when you go to market. The distribution system you choose will tell you what percentage figures to plug into your calculations for things like markup from wholesale to retail, sales commissions, shipping costs, and other factors.

There are quite a number of different distribution systems, and they all have their own peculiar effect on those calculations of yours. The next chapter in this book deals with all that in detail. You'll need to read and reread that chapter before you can actually refine your wholesale price, but for right now you just need to track down a little information.[1]

In order to figure out your best method of distribution, you'll have to go to the kind of stores that will be likely to sell your product when you get it on the market. Let's say you've got some kind of automotive accessory in mind. You go to your friendly local accessory store and buttonhole the manager. Don't be shy. After all, how many times a day does the owner or manager of a small store have a chance to become a marketing expert and pass out learned advice? He'll usually be happy to spend more time with you than you're interested in spending with him. But then, most things in life are a trade-off.

What you want to learn from the retailer is where his store gets products like the one you have in mind. It couldn't hurt to find out how much he marks up products of that type, either, although his distribution system will probably define that markup. If you're nervous about letting people know precisely what your idea is at this point, find something in the store to use as an example. Something in the same general price range, aimed at the same general type of consumer as your product. Level with the guy and tell him

---

1 As a matter of fact, you really ought to finish reading this whole book before you actually set out to market your invention. I've tried to put this thing in some sort of logical order, but there's a lot of information to cover, and much of it interlocks. To help you organize it all, I've included an inventor's checklist at the back of the book as Appendix I. It puts all your chores down in order, letting you check them off as you go along.

you can't reveal exactly what the product is yet, but you're going to be marketing a new product, and you need some help. If you look young enough, tell them you're a student. That always gets sympathetic response.

Once you know where he gets things of that nature and what his markup is, you can go to *his* source—a wholesaler or distributor, manufacturer's rep, or whatever. Find out what the wholesaler's markup is, what the distribution system is, whether or not there are other means of distributing the same sort of product. Find out what the advantages and disadvantages of any alternate methods are and figure out which of them makes the most sense for your product. Do it all more than once, just to be sure you're not getting funny information from any of your marketing experts.

KENTUCKY WINDAGE

Now you've got the means to refine your product costs and arrive at a fairly solid wholesale price. You know the markup structure you'll be working with, and you have a pretty fair notion of the other costs involved in the distribution system (if you've read the next chapter, that is). You've got a number you can rely on, more or less. You'll still find that you may have to use some Kentucky windage in some of your calculations, so don't get panicky if you can't pin down a rock-solid figure. When you err, try to err a little on the high side. I once made a horrible mistake at this point in the development of a new product. It cost me more of my own personal, private dollars than I care to remember.

The product was a doll, made of felt and called (for reasons that seemed to make sense at the time) "Little Lumpsie." I went out and collected bids from fabric suppliers, from sewing factories, accessory suppliers, the whole bit. I was really proud of myself. "Nothing to it for a professional like you, Kracke," I said to myself. I didn't realize that my experience in printing didn't necessarily make me a genius in the field of soft goods.

Anyway, the doll costed out beautifully. I ran off a couple of samples and it was presented to FAO Schwarz, the ritzy toy store in New York. The company loved it. Placed an order. Hot damn, I was in the doll business!

Boy, was I in the doll business! All those suppliers who gave me firm bids turned out to have been doing a little fibbing. When I went into production, they doubled their prices. All of them. Every last one. It was like a giant conspiracy to drive me to the wall.

But I was committed to fill the order, so I paid the price. I didn't want to renege on a deal with FAO Schwarz because I'd like to be able to come back to them again sometime with another idea. I'd rather take my beating than lose a good contact. I lost money on every last one of those damned dolls. Worse yet, it was a success. Schwarz reordered again and again. At that price, I didn't need the volume.

Ultimately, I sold the idea to Fisher-Price Toys for a royalty. They were able to scale the idea down a little, cut their costs and sell the toy at a profit. They sold a lot of dolls and I got back my loss (at $.05 each, royalty). I chalked up another lesson.

The lesson was this: When you're getting bids from suppliers, don't be afraid to ask them for a contract. Put in penalties they have to pay if they don't deliver on time or at the price they quoted. I'd guess something like 10 percent is enough leeway to allow them in their price. They won't be overjoyed, but it will make them think twice about shooting you a lowball bid.

As a beginner with a new idea, you may feel you have to go around to suppliers with your hat in your hand, especially when you're talking about fairly small orders. Don't worry about it. It's *your* neck that's being risked. It's *your* money they're playing with. You deserve all the protection you can get. And if the supplier is a smart businessman, he'll be willing to invest a little understanding at the beginning in order to reap a little loyalty later on. We had a perfect example of that kind of thinking when we first ordered some small plastic bags for our package of Rickie Tickie Stickies.

PRETEND YOU'RE for REAL

I called a number of companies and asked them about sizes in stock and how we might do business. With one exception, the companies were just not too interested in dealing with a start-up company who wanted a tiny order of small bags. Their prices reflected their reluctance.

One man, however, whose business was miles and miles from our home, was amazingly cooperative. His company was one of the biggest, we later found out, so he wasn't exactly desperate for business. It's just that he was decent and smart. He listened patiently to my needs and, most surprising of all, shipped our order out to the house with nothing more than a promise to pay when the bill arrived.

As it turned out, his trusting response to that first order paid off for him and his company. As our operation grew, we kept ordering all our bags from him. At first the orders were pretty puny. Later they became respectable enough to attract attention from his competitors. Suddenly the bag makers who didn't have time for our silly little company began dropping by to see us, offering special deals. Weighing the one-time special deal against the benefits of a long-term relationship with a conscientious supplier, we stayed with our original source. We never regretted it, and neither did he. Altogether we bought several million bags from him by the time we got out of the flower business. And the next time I need bags, I know the first number I'm going to call.

Anyway, don't be afraid to act like a real businessman when you're getting bids. Even with a contract, you have no guarantee you won't be hustled by some unscrupulous supplier. But giving yourself at least that much protection lets people know you mean business. Better that than to go blithely off to market on the basis of a loose verbal quote. That way could be disaster. And insolvency.

THE OLD WET-FINGER TRICK

So now you've got a wholesale price, and you know the distribution system you'll probably be using. Put them together and, *voilà,* the magic number emerges: your retail price.

It's probably too high.

It almost always is. So now you've got to go back into your costs and find ways to cut them. Find different suppliers. Find different ways of manufacturing and packaging. Rethink your idea. Bring your wholesale price down to a point where it leaves you with a viable retail price.

But how can you tell what's viable and what's not? There are ways, ranging from the ridiculous—our efforts at the beginning of the Rickie Tickie Stickies adventure—to the sublime—the kind of market research done by the industrial giants. Somewhere between those extremes lies your answer.

In our case, with the Stickies, we sort of held up a wet finger and decided that $.25 a Stickie was about right. It fit our estimates of manufacturing cost and left us a decent profit. Then, when the neighborhood kids came around asking to buy the flowers, they confirmed the fact that we were correct in figuring on two bits as the retail price.

Another way is to go around to the stores and see what similar products are selling for. Even if there's nothing similar on the market, you can get some sort of feeling for where your product will fit into the general scheme of things in that line of goods.

Finally, there's a somewhat more sophisticated, although still fairly straightforward, way to find out if your product will sell at your price: try selling it. On a small scale, of course. There are a number of advantages to doing this. First, of course, it will tell you what you want to know about your price. More important, in the long run, is the fact that it gives your product a sales track record in the real world. That's extremely important to you as a means of attracting attention when you try to sell your idea to some manufacturer. It's equally important to you in getting into new, bigger, more lucrative markets with your product. If you can prove you have a winner in a limited yet projectable market sample, you have a powerful argument in your favor. Needless to say, it could help you when you have to go looking for financial backing.

When we were able to attract the attention of an industrial giant who wanted to buy the rights to produce our Stickies, it was our solid, real-world sales figures that opened the door. Those numbers made it possible for our little gnat of a business to be noticed by the great, powerful moose of industry. We were manufacturing the product. People were buying it, stores were reordering it. The gnat bit, and the moose scratched. It was a triumph of market research.

At the time, of course, we didn't think of it as market research. We were just out there selling the Stickies the best way we knew how. We were hitting the neighborhood stores and a few department stores, order book in hand. We had run off that short sample run that became our opening inventory. From there on, we just kept half a jump ahead of the orders as we went along. Well, maybe sometimes we were half a jump behind. But the theory was correct. We simply spent as little as we felt we could get by with to keep up with sales in our rather limited sort of market area. And as reorders climbed, our number of stores expanded.

Reorders, by the way, are called "turns at retail" (a little something to liven up your cocktail party chatter). A product is said to "turn" at retail when it is reordered by a store. The measure of a product's ability to perform is the number of turns in a given period of time. In our case, it didn't take a mathematical genius to project our sales potential.

Sales in the five hundred Southern California stores we finally reached had been going briskly through October, November, and December of 1967. (We were happily unaware, at the time, that nobody buys wholesale in November and nobody, but *nobody* ever places a wholesale order in December. We merely noticed a worrisome little lag in sales those two months. Blissful ignorance.)

There were approximately 24,500 similar stores in the United States, according to some high-powered research we did at the local library branch.

And "average" product in those stores turned at retail about four times a year. Rickie Tickie Stickies were not just turning, they were positively spinning. Something between twelve and eighteen times a year.

In making our sales projection, we ground in one more factor. Southern California is considerably more receptive to new ideas than most parts of the country. People here are a lot more likely to buy something new and strange, which is maybe a commentary on Southern California. Therefore, we arbitrarily cut the national sales projection in half. Simple arithmetic, based on a fair sales test in a California sample, indicated a one year, $18 million potential at wholesale across the country. I settled for $9 million in my presentations. As it turned out, the number was over $10 million.

You ought to be able to do pretty much the same sort of calculation with your product. Figure out what kind of sample market you can deal with (bigger is better, more projectable), crank up limited production, then get out there and sell. And keep track of your turns at retail. Once you have a reasonably fair idea of how the sales are going, you've got a powerful attention getter for your product. You'll also know whether or not your price is right.

You might even be able to get some stores to cooperate with you in a little market research gimmick. Try selling the product in one store at a low price, while some distance away you're selling the product at a high price. Meanwhile, sell your product somewhere else at the regular price. Try to match the stores as closely as possible and try to make sure they're far enough away that they won't have any of those same customers. Watch what happens to your sales at those stores. If your rate of sales at the high price is about the same as the rate at the regular price, you can obviously afford to charge more for your product and make more of a profit for yourself and for the stores that carry your product. If sales surge markedly at the low price, well—better find a way to cut costs.

One hidden advantage of test marketing a product is that you may be able to use your successful sales efforts as a lever to force distribution. I'm involved right now in a classic example of that.

DEAFENING SILENCE

The product is a book called *The International Handbook of Jockstraps.* (I didn't say it was a high-tone product, did I?) It's a gag book, a whole series of cartoons of different kinds of jockstraps. The "Jock Cousteau," for example, is a picture of a jockstrap with a diving mask on it. Some of the cartoons are funnier than others, but by and large it's a fairly good example of the sort of gag gift item you'll find in a lot of stores.

When I took it around to publishers, however, the silence was deafening. None of them believed in it. I couldn't get it sold anywhere. So what to do? I didn't want to spend a lot of my own money getting it published. And I didn't want to give up on it. What I *did* want to do was prove to some publisher that the book would sell and would be a profitable item.

So I did something that I've done before. I co-ventured the book. I made a deal with a printer giving him a percentage of the book in return for printing it free. I'm out nothing but my time and the time of my fellow authors. The printer has run off the book during downtime on his presses, so he's out essentially nothing. And we are, at the moment, into our tenth printing. We've sold over 210,000 books in a limited number of stores, distributing the book through gift and stationery reps.

We're just about at the point now where I feel confident about going to a publisher, showing him my sales figures, and asking again if he's interested in publishing the book. I think he'll do it this time.

Those, then, are the main areas that come under the general heading of doing your homework. You'll find plenty of other jobs to be done as you develop your idea, of course. You're going to work very hard before your reach your goal. And you'll find doing your homework is such a recurring theme in this book that you'll be tired of hearing it. But I have to tell you it is awfully important to you.

Yes, it *is* possible to sell an idea without doing your homework. It is also possible to fill an inside straight.

But I wouldn't bet on either one of 'em.

Which Way TO Go

Many roads lead to the marketplace, but some are better than others.

IT'S TIME TO START THINKING ABOUT MAKING MONEY. THE QUESTION before the house is this: How will you go to market with your product? It's a question you can't put off any longer. You *must* get out into the marketplace one way or another if you want to make money. And there is an endless variety of ways to go. So which will it be?

Will you make it and sell it yourself? Sell the idea to a manufacturer for a royalty on his output? Build a mail-order business operating out of your home? Sell the product to a Major Catalog Marketer? Sell it to one of the Specialty Catalog outfits? Maybe form a company to make the product, then sell the company? Turn the product into a premium or sales-incentive item for some company like a fast-food chain or a soft-drink bottler? How about setting up your own Web site and joining the e-commerce set? Or what?

Before this chapter is over, we'll have a good, hard look at all the options. We'll find out the ins and outs, the advantages and disadvantages of all the main routes to the retail market. But let's get one thing straight right up front. There's only one way you *want* to go.

ADVANCE AGAINST ROYALTY? YES!

What you want to do—desperately—is to sell your idea to a manufacturer for an advance against royalty. Anything else is second best. A fall-back position.

Collecting royalties is the cleanest, neatest, most profitable, simplest way for you to make your fortune in the inventing game. Naturally, it works out that the most difficult and frustrating task you'll face in this business is making this type of sale to a manufacturer. You're probably going to have a lot of doors slammed in your face. However, all it takes is one "yes" to make it all worthwhile.

Usually at about this point, people begin to wonder if it could possibly be worth all that aggravation. The answer is yes. Unequivocally yes. Absolutely yes.

"Well, yeah, but isn't that giving away a big piece of the profit on your product?" people often ask. "Wouldn't you make more money if you manufactured and sold it yourself?"

Theoretically, yes. In practice, no. To sort out the objections, let's take a look at what's involved in going it all on your own. As a matter of fact, you may have to do that for a while anyway, to establish enough credibility in the marketplace to make a royalty deal. It can be good for you in limited quantities, as it gives you a feel for your market, points out some of your problems while there's still some hope of solving them, and tells you how successful your product is likely to be. And as I said earlier, a little successful sales history is a strong point in your favor when you're negotiating with a manufacturer later on. So you might as well know what going it alone is all about.

GOING it ALONE

In its simplest form (a misnomer if I ever heard one), going it alone starts with devising a way to manufacture the product, then working out packaging, pricing, and merchandising. The next step is to manufacture a quantity of the product, load up the family car, and start schlepping your product around to a lot of stores, making sales calls, and delivering orders. Later, after you've established a little headway, the smart move is to get a sales rep or a network of sales reps and expand to a regional operation. Maybe eventually to a national operation, if it comes to that.

In a way this is all kind of appealing. With you as head honcho of the operation, all the profit there is to be made rolls merrily into your bank account. As top banana of the business, you have total control of how the product is made, how it's merchandised, how it's distributed, where it's sold, and how much it sells for. The whole works.

However, being the boss isn't all hearts and flowers and deposit slips. I soon discovered, when I started out with the Stickies, that getting all the control and all the profit means you get all the headaches, too. It's a package deal. And some of those headaches can be real migraines.

For instance, picture yourself doing the old soft-shoe routine for some irate customer whose order won't be filled on time (the guy who makes the boxes for you missed his delivery deadline). I got to be a regular Fred Astaire when I was selling the Stickies.

Or how about conjuring up a mental image of your living room or garage, all stacked to the rafters with a few hundred gross of big cardboard cartons full of your product, due to a slight misunderstanding with the shipper.

Imagine you and your banker on a first-name basis—yours, not his—because your loans are so big that if you go, he goes. You begin to get the picture.

And that brings us to the question of who pays for all this manufacturing and merchandising activity. Somebody has to scrape up the money for tooling, for the design and printing of packages and store-display units, and for shipping. Somebody has to buy the raw materials, cash up front. Somebody has to pay the help and absorb all the costs of being in business— everything from phone bills to computer repairs. Somebody's got to spend some big money.

Guess who.

Okay, so there's a lot of work, expense, and worry involved in manufacturing and distributing your own product. But that's not the end of your problems. There's an even more insidious danger you face when you go it on your own. You could be setting up your idea to be stolen.

DISTRIBUTION is EVERYTHING

You are limited in the number of outlets you can reach. (How many stores can you call on with the family station wagon?) But there are companies out there with big, efficient, nationwide sales and distribution organizations just shopping around for new ideas to steal. Yours, for instance.

So there you are, selling briskly, making a nice profit out of the back of your car as you call on your handful of stores. Maybe you've even built up a nice regional business with the beginnings of a fairly decent distribution system. Then along comes the big, unscrupulous manufacturer who notices you. He makes a few discreet inquiries. Yep, your product looks like a winner, all right. And while you're trying to figure out how to reach another half-dozen stores, he's cranking up the factory, filling his distribution pipeline with a copy of your idea. Before you know what hit you, he's saturated the market with a well-oiled national blitz.

You're left with the remains of your back-room business and your headaches while he's off scouting up another new idea to knock off. After all the legal niceties are over, after all is said and done, the guy who gets there biggest and fastest in the national market is the guy who gets most of the marbles. I've told hundreds of inventors. Now I'm telling you:

An idea is *one thing.* Distribution is *everything.*

The way to avoid this problem if you're going it alone, then, is to get maximum distribution rolling just as fast as you can. That means getting a manufacturer's rep or a network of reps. A manufacturer's rep is a salesman who calls on a regular list of retail outlets and/or distributors in his territory and sells them the product of various manufacturers he represents. Hence the term. The rep goes out and sells an order for a couple dozen gross of your product to a store or distributor and sends the order on to you. You ship it out to his customer from the warehouse or from your garage or wherever you're storing the goods. And you enclose an invoice. Simple as that.

You may connect with a rep who has national coverage. If not, have him help you put together a national network of similar regional reps. He'll know who the good ones are and be able to supply arcane bits of knowledge, like the fact that whatshisname in Atlanta sells like gangbusters for the first six months with a new product, then fades in the stretch. Or whatever.

Needless to say there's more to it than just picking a rep out of the Yellow Pages, calling him up, and signing on. He's in business to make money, the same as you are. So you're going to have to convince him that you will be a profitable addition to his list of clients. Sales figures can help, if you've been out peddling the product on your own for a while. It's great to be able to show him where you've been selling, how much you've been selling, and how often your customers have been reordering. If you haven't been selling yet, you'll have to make a pretty convincing case for your sales estimates and projections. Bad reps jump at fantasy. Good ones react to facts.

The rep will be interested in other things, too. For instance, he'll want to be sure you will keep a big enough inventory on hand to fill his orders when they come in and that your rate of manufacture can keep up with his rate of sales. Finally, you'll have to convince him that you're solid—that you're going to be in business for a while, especially under the pressure of growing national sales. (That sounded absurd to me when I first heard it, but I soon learned that keeping up with a spurt of sales can put awesome strains on your finances. Prosperity can be a killer.)

94

There's one more thing he's going to want to see: a production sample of your product. Not a handmade model. Not a picture. Not a glowing, animated description with arm waving and fast talking. A real, honest-to-God production sample, just like the ones he'll be selling for you. There are times when you can get away with a model or picture, as I mentioned earlier in this book, but not when you get to this stage. Besides, he's probably going to be displaying your product in his showrooms and in various tradeshows around the country where his customers come to order their wares for the upcoming season. If he doesn't have the product, he's not going to get many orders.

So much for what the rep expects from you. You have a right to expect a few things from him, too. Like Will Rogers, I've never met a rep I didn't like. But I have to point out that they're all a little optimistic when it comes to the territories they claim to cover. Make sure the ones you use really do cover the territory they claim or you could miss out on important sales.

When negotiating with any manufacturers' reps, start out with the premise that they need to have an office and/or showroom space and at least one salesman and a secretary in each major metropolitan area in each state in which they want to represent you. If they lack coverage in some areas, find reps who do cover these places and give them that part of the territory. The first rep will kick and scream, but you're the boss. And you'll be right.

For your information, there are twenty-one major metropolitan marketing areas in the United States. You should be represented by real, living people—preferably with showrooms—in each one of those areas.

Normally, you'll assign territories to reps on the basis of geographical regions, usually divided along state lines. They'll ask for exclusive rights to sell your product within the agreed boundaries, and that's fair. Up to a point. And that point is, you may lose some good sales if you don't limit your reps to "vertical" lines of distribution.

To illustrate: Ever stop to consider how many things are sold in military PX and Commissary Stores? Jillions! And how about school bookstores—high schools, junior colleges, colleges, universities? Now you're beginning to catch on. Who sells those places? Reps. The same reps we have been talking about? Normally not. These guys are specialists. Have you ever tried to figure out a government form on anything? You should see their order forms! It takes a specialist, believe me, to understand them. So keep that in mind when you're assigning reps. You're perfectly within your rights to limit your reps both geographically and vertically by retail sales categories.

How do you find school reps and military reps? Call your local college bookstore and ask for names. Call your favorite military base, ask for the PX (or BX in the Air Force, Navy, and Marine Corps) and do the same thing. Once you've selected a rep in your area, he or she will lead you to other such specialist reps around the country.

One last word of caution. You'll be approached, if you're successful in a regional or local test, by someone claiming that the only way to go is with a national distribution organization. He may be right. There are a lot of advantages. Just remember to find out if he has people and showrooms in all of the twenty-one major metropolitan areas of the country.

¿DONDE ESTA EL STICKIE?

Foreign sales can be a very important part of your success story, and they're not as hard to get as you might think. The biggest single order we ever got for the Stickies was $52,000 worth at one time. I've still got a Photostat of the check on my wall. Wish I still had the real check. The buyer? Ahlens department store in Stockholm, Sweden. And when those Swedes make up their mind to sell something, they don't fool around. They built a whole special promotion around the Rickie Tickie Stickies and sold out all $52,000 worth in one month. That was $104,000 at retail!

Breaking into the Swedish market was easy, too. In the first place when we started out and I was patenting and registering and copyrighting and things like that, I registered the trademark in a whole bunch of likely sounding countries, including Sweden. When things were moving nicely here, I called Scandinavian Air Service (SAS) and asked for some information about marketing in Sweden. The SAS lady came to see me and gave me all kinds of detailed information about the Swedish market, how to enter it, what regulations there were, what the marketing setup was, and so forth. She had it all down, chapter and verse, and she knew what she was talking about.

By contacting someone like SAS or various countries' trade attachés or their industrial development offices, you can find out just about everything you need to know to market your product in foreign countries. They will also lead you to foreign reps or brokers.

There are more advantages than meet the eye, too. For instance, even as the enormous enthusiasm for Stickies was wearing down here in the United States, new countries were just discovering it. By rolling out into new countries, you can keep momentum going for your product, keep your manufacturing operation at a relatively steady level, keep your sales curve more like a curve than an alpine range. And, of course, checking up on your sales overseas is a pretty snappy excuse to do a little world traveling. "Ah, yes. Off to Rio to check up on business."

As I hope you can see, going out there on your own can be difficult. That's not to say it can't be done, though, if you conclude it's the best way for you to reach the retail market. Before you come to that conclusion, let's take a look at some of the alternatives. Which brings us back, after all our digression, to the ideal way to go: selling your idea to a manufacturer for royalty.

THE only WAY to FLY

The biggest advantage of going this way is that it puts your product on the market through the manufacturer's distribution system—hopefully, a big, complex, efficient, nationwide one. Where you would have had to sweat and strain to work your way up to national sales and distribution, the manufacturer can jump right into the market with his best shot immediately. It gives him—and your product—a thumping big start on knock-off artists.

For another thing, it puts the responsibility for financing the business on the manufacturer's shoulders, not yours. You don't have to spend any of the money for tooling. You don't have to bear the expense of being in business. Best of all, you can dump all the worries on someone else's desk.

The first step to take is to pick out a likely manufacturer. That's not as simple as it sounds. One of the most common mistakes made by inventors at this stage of the game is choosing the wrong manufacturer. Far too many inventors try to sell their idea to a company whose production capabilities don't match the product, whose market is totally wrong for the product, or who is simply not prepared for one reason or another to deal with the inventor's idea.

A friend at Fisher-Price told me that outside inventors bring them some 2,000 ideas a year. He has to reject about 1,400 of them out of hand because they don't match his product line—they can't be made by the equipment in Fisher-Price's factory, they don't fit into their product categories, or they are totally unsuited to Fisher-Price's market. All 1,400 of those poor inventors could have saved themselves a lot of trouble by simply going to a toy store and looking around for fifteen minutes.

So the cardinal rule for choosing a manufacturer is to find one that makes and sells products compatible with yours. Look on the shelves of the stores where you expect your product to be sold. See where the manufacturer you have in mind fits into the picture. If you have an idea for a product made out of wood, there's no point in approaching a company that makes everything out of plastic. If you have a product that ought to be sold in hardware stores, don't go to a manufacturer who sells most of his stuff in department stores. Check the price, materials, and market of your product against the price, materials, and market of the company you're planning to approach. If they don't match, don't bother.

Something else to bear in mind is that you can get into an immature, unsophisticated industry much more easily than into a mature one. The classic example is toys. Toys are a very, very difficult market to break into with a new idea. The industry is dominated by a few highly sophisticated, extremely rich companies. The competition is intense. Cutthroat, even. Almost no outsider ever makes it with a new toy idea. My own experience of selling my Little Lumpsie doll to Fisher-Price a few years ago was an incredible exception. Bob Hicks, who bought the idea from me, told me that he has seen 50,000 new-product ideas from outsiders in his twenty-five years with the company. In all that time, out of all those ideas, he's bought five, including my Little Lumpsie.

On the other hand, the gift and stationery field is immature. An outsider has as good a chance as anybody in cracking the gift and stationery market with a new idea—and Rickie Tickie Stickies is the proof of that particular pudding. There's really only one, big dominant manufacturer in the field, and that's Enesco out of Chicago. Gift and stationery stores themselves tend toward mom-and-pop operations, small stores where the owner waits on customers. There are a few big chains. So a neat, marketable idea—even something as off-the-wall as my Stickies—has a good chance of making it.

Part of the reason for the difficulty of cracking a mature industry is that the biggies have such heavy investment in their own idea-generating organizations that they don't *need* any outsiders. In fact, after seven months of politely trying, I couldn't get into Whamo Corporation (Hula-Hoop, Frisbee) to discuss this book. They were *that* busy. Or something. A vice president at a major toy manufacturer explained to me that a development group of fifty people will be involved in coming up with a new idea within the company. And that's just to get it to the point where it's ready to be refined by another development group of from 200 to 250 people. With that kind of high-powered idea organization, it's small wonder that fewer than one half of one percent of the ideas that company carries to some stage of development come from outside the company. And most, by far, of those outside ideas come from a few professional inventors with whom they regularly deal. The odds against you or me getting to first base in that kind of situation are astronomical.

In addition, the biggies in a mature industry have researched the field so intensely and have gained so much experience that they know more than any outsider could possibly hope to know about their market. Fisher-Price, for instance, won't hesitate a moment to mount a $100,000 research program for a single toy idea they plan to market. Granted, as my friend points out, that research is far from a science, but it sure beats holding up a wet finger and making a wild guess.

Most important of all, an idea has to be capable of generating enormous profits before it's worth bothering with by one of the giants in a mature industry. It has to support 250-person development teams and $100,000 research projects. At Fisher-Price, I'd guess that an idea would have to look like $10–20 million in sales a year at wholesale before they would be interested. Anything less than that wouldn't be worth the time, effort, and investment that would go into it. It could be a swell idea. A perfectly saleable idea. It could tickle the fancies of Fisher-Price and Mattel, but unless it can generate big, really big money, they won't buy it.

All of this boils down to a couple of rules for you to bear in mind when you're setting out to sell your idea to a manufacturer: Make sure your product matches the company's needs and capabilities. And try to stay away from the ideas of mature, sophisticated industries like toys or games or clothing, to name a few. (Don't throw away a good idea, just because it happens to be a toy, mind you. Just be aware of the odds.)

MORE than YOU WANT to KNOW

Every route to the marketplace involves its own special distribution system. Since every distribution system has its own peculiar set of costs of distribution, you ought to understand the markup structure inherent in each. You'll need that information to arrive at your wholesale and retail price, as we discussed in the last chapter. To simplify matters, we'll use a hypothetical product that sells at retail for $2.00 to explain each distribution system. Every time we come to a major distribution system in this chapter, we'll stop and take a look at what happens to that $2.00 item, so you have a benchmark to work with.

Whether you go it on your own or sell your idea to a manufacturer, you'll probably be involved in one of the two most basic distribution systems: the distributor/wholesaler system or the factory-direct-to-retailer system. We'll get to how that $2.00 is spread around in each of these two distribution systems in a moment.

At this point you're probably wondering why in the world everything isn't distributed on a factory-direct-to-the-retailer basis rather than through a distributor. After all, the manufacturer gets more money going direct. Good question. Fortunately, I have a good answer. Consider the basic distributor/wholesaler system:

BASIC DISTRIBUTOR/WHOLESALER SYSTEM

CONSUMER
⇑
RETAILER
⇑
REGIONAL OR NATIONAL
DISTRIBUTOR/WHOLESALER
⇑
MANUFACTURER'S SALES
REPS OR OWN SALES FORCE
⇑
MANUFACTURER

Who gets how much on a typical $2.00 retail item with this form of distribution?

The consumer pays the retailer (plus tax) ....$	2.00
The retailer pays the distributor	1.20
The distributor pays the manufacturer	.90
The manufacturer pays his sales force	.07
The manufacturer allows cash discounts and freight credits amounting to about	.05
The manufacturer's factory gross is	.78

Now consider the factory-direct-to-retailer system:

FACTORY DIRECT TO RETAILER

CONSUMER
⇧
RETAILER
⇧
MANUFACTURER'S SALES
REPS OR OWN SALES FORCE
⇧
MANUFACTURER

Who gets how much on a typical $2.00 retail item with this form of distribution?

The consumer pays the retailer (plus tax).....................$ 2.00

The retailer pays the manufacturer..................................   1.00

The manufacturer pays his sales force 10

(This will vary from $.05 to $.20 depending on
the industry. An average of $.10 is fair.)

The manufacturer allows cash discounts
and freight credits amounting to about 05

The manufacturer's factory gross is 85

A good Class-A distributor earns his money. He warehouses the product. He sells the product using *and* paying his own sales force. He bills and collects the money. He handles the problems of damage, returns, and the like for the manufacturer. And probably the most important thing he does is to pay the manufacturer promptly. Not because he's just nice. That 2 percent discount for prompt payment does add up.

A good distributor network can save a manufacturer other problems too. Fifty Class-A regional distributors can cover the United States for just about any product you could invent, regardless of what kind of store the consumer finds it in. Those distributors will be reaching upwards of ten thousand separate accounts, with all the attendant problems of billing, hand holding, and collecting. The manufacturer has only fifty regional distributors. Fifty bills at the end of the month instead of ten thousand. That's a good beginning.

Unfortunately, all industries don't have good national distributor networks. Gift and stationery, for example, is one. In the heyday of Rickie Tickie Stickies, I would have been happier to have had fifty large accounts instead of the 6,800 little ones we ended up with. And the bottom line would have come out about the same. It costs money to do it yourself in the distributing end of the business.

MAIL-ORDER BUSINESS

If you haven't been able to sell your idea to a manufacturer, and you don't want to mess around with making sales calls, and you're not ready to sign up with a manufacturer's rep, there's another way to go. It has some impressive advantages. It's called mail order.

By setting up your own mail-order business, you give yourself broad market penetration immediately. You can reach anywhere in the country—anywhere in the world, for that matter—with your ads. You can become an instant national or international company, not limited by the number of stores you call on in person. Your customers are everywhere.

You can even do a mail-order business while simultaneously doing retail business in stores. Be careful, though. Quite a few retailers—and quite a few reps, too—don't take kindly to competing with mail-order sales of the same product. It works the other way around, too. Retail sales can kill off a mail-order business. A customer may feel better about getting the product immediately at the store than waiting around for the mailman to deliver it. And some products simply aren't as good in mail-order business as they are in a store.

This sales segment is way too complex to address in this chapter. You would be well served to get on the Internet or go to the library for research if this way appeals to you.

There isn't any nice, clean way to show what happens to our hypothetical $2.00 item in the mail-order distribution system. That's because there are several variables involved. It works this way:

CONSUMER

⇧

MANUFACTURER

The consumer pays the manufacturer direct..................$ 2.00

But before the manufacturer can arrive at his factory gross, he has to deduct the cost of advertising and shipping involved in that $2.00 sale. Amortization, in other words. He has to total up all the advertising expense that led to a given batch of sales, then divide it by the number of sales those ads led to. (If he spent $500 on ads and generated 1,000 sales, his advertising expense is $.50 per sale.) Then the manufacturer deducts the cost of shipping the order. The result, after all that is done, is the factory gross.

Getting into the mail-order market doesn't necessarily mean you have to do it yourself, though. There's another way to sell by mail. A much better way, which eliminates most of the drawbacks we just discussed. You sell your product to a catalog marketer. If your product is something that interests such merchants, you could have a one-way ticket to fat city. You'll be plugged right into their national catalog sales and distribution systems, to say nothing of all their retail outlets. Think Williams-Sonoma. That's better than having the hottest rep in the country, believe me.

EASIER SAID than DONE

It's easier said than done, naturally. You don't just stroll in and make your sale to the biggies just like that. You have to convince them that your product will make a nice profit for them.

If that's beginning to sound like a recurring theme, you're right. And so is the routine you have to go through to make the sale. You have to do the same things you do if you want to convince a rep to take on your product. Show sales figures, if you have them. Show solvency and reliability. Show the ability to keep up your supply with their demand. And you have to show a production sample, not a model or a drawing.

In many ways, selling your product to one of the major catalog marketers is almost as good as making a royalty deal with some manufacturer. For one thing, it lightens the burden of doing business, because making one big sale is a lot more profitable and painless than making a dozen small sales. Even shipping will be cheaper, since you'll be sending a big batch at one time to the marketer's warehouses. All that adds up to a more profitable and lower-priced product, which means more money in your pocket.

One more thing, before we leave the topic. Often you will be asked to give the catalog house an exclusive right to carry your product. That is, they will ask you not to sell it through any other outlets, at least for a period of time. Usually twelve months. That's to give them the chance to establish the market for the product, skim the easy sales and take the position as the innovator—the company that brought this great new item to the consumer. Later, when they may no longer have an exclusive, they hope that the public will still think of them first when they think of your product.

Naturally, it's to your advantage to avoid giving anyone an exclusive. It just cuts down on the number of places you could be making sales. But you may have no choice in the final negotiations. It's worth holding out for, but it's not worth losing the big sale for.

Oddly enough, sometimes a major catalog retailer will do a complete about-face and actually help you market the product someplace else rather than ask for an exclusive. A case in point was a new kind of spillproof ice-cube tray that a national chain decided to develop. Now, nobody gets rich selling ice-cube trays. Every refrigerator comes with a couple of them, and there's just not that much of a market for replacements. Still, it was a good item, and if someone else could stir up an awareness of the need for spillproof ice-cube trays, quite a few of them could be sold. So the catalog retailer helped the manufacturer sell the tray idea to a big appliance company. The company advertised the clever trays as a product advantage in its line of refrigerators, making the spillproof trays well known. Then the catalog retailer advertised the trays in its catalog, saying that you could have this new product feature for your old refrigerator. It sold a lot of ice cube trays. The catalog retailer got what it wanted, and the appliance company had a new benefit to advertise. Everybody won, with a little help from the cataloger.

There's a whole world of specialty catalog companies out there, reaching an entirely new spectrum of consumers with a completely different kind of product. They represent another way for you, as manufacturer of your product, to get at your market. We're talking about companies like Spencer Gifts. As a general rule, the products they sell are novelty items, usually of fairly low price. The pieces are usually relatively compact and easy to ship.

The big catalog companies we were discussing earlier have a tradition of carrying big, bulky items, like complete car engines and kitchen stoves. They're prepared to deal with the shipping problems involved. The specialty-catalog people aren't fascinated by that kind of challenge.

In many ways, the specialty catalog is tailor-made for inventors like you and me. So many of the things we come up with seem to be the type of thing you find in these books. More important, the people who put out those catalogs are constantly searching for people like us—people with novel ideas they want to develop. The big, traditional catalog marketers are interested in new ideas, too, but they also lean heavily on regular run-of-the-mill merchandise for their sales. The specialty marketer depends entirely on unique products, special new ideas. Hence, the term *specialty catalog*.

Here's the catalog model:

CONSUMER

⇧

CATALOG

DISTRIBUTOR

⇧

MANUFACTURER[2]

This consumer pays the catalog distributor ....$ 2.00

The catalog distributor pays the manufacturer...............   1.00

The manufacturer's factory gross is 1.00

The factory gross will vary up to plus or minus 10 percent because of two important factors. Chains normally represent large-volume sales and, therefore, they qualify for large-quantity discounts. They will also normally receive the maximum freight allowances (like for free) and will take advantage of all cash discounts.

When I noted "qualifying for large-quantity discounts," it reminded me to remind you of a very important fact if you end up being a manufacturer. Decide *before you go to market* if you will be offering quantity discounts.

Also, anticipate discounts for *foreign* brokers before you finalize your numbers. Double check these quantity discounts, then check them again to make sure large sales don't fall into the "feeding the monkey to watch him crap" category. If you're satisfied that you can make a profit while offering quantity discounts, then *publish* the fact. Even if it's only a few copies. Have some dated copies in your files. And be sure you offer the *same* terms to everyone. *All the time*. The Feds take a dim view of your offering different strokes to different folks, especially if you end up selling overseas at different prices than here. They call that "dumping," and if you can't prove it is a consistent sales policy, they can really dump on you!

---

2 This type of sale is normally a "house" account in which a principal of the manufacturing company is the sales contact. It's that important. There are reps specializing in chain sales. When involved, they usually get around 10 percent.

Now we'll look at a "specialty" deal. The same distribution chart applies:

CONSUMER
⇧
CATALOG
DISTRIBUTOR
⇧
MANUFACTURER

Here's where we get a variation on the theme:

This consumer pays the catalog distributor ....$ 2.00

The catalog distributor pays the manufacturer...............   .50

The manufacturer's factory gross is 50

Why? As I said, the specialty catalog has basically the same start-up costs of a major catalog. Prorated against much less dollar volume per item, the difference has to come from somewhere. That somewhere is usually the hide of the manufacturer!

A PREMIUM PREMIUM

There's one way to sell your product I'll bet you never thought of. You could sell it to some company as a premium item, the kind of thing that's given away or sold at a low price when you buy the jumbo-juicyburger and fries at your local hamburger hustler. The kind of thing you send a box top and a buck for. That may strike you as a kind of goofy way to get sales, but quite a few products get to the public that way. It could be either a jumping-off place for you—a way to get some exposure before you go to retail—or it could be the natural market for your product.

You don't have to be the manufacturer to sell your product as a premium (although you make more money if you are.) If you can meet the heavy production demands of the premium users, fine. But if not, most companies who buy premiums have a regular stable of manufacturers with whom they work. They'd be happy to buy your idea from you, either for a straight cash buy-out or—more rarely—on a royalty basis.

The price range of premium items varies vastly. Some companies, for instance, have sold everything from sailboats to binoculars as premium items.

The big-ticket or moderate-ticket items used by those companies represent one end of the premium spectrum. The other end is represented by the little inexpensive gimcracks that are handed out with a purchase of the promoter's product. In between, there's a vast area of premium users who should also be considered as possible markets for your product.

Sunkist, the citrus giant, uses some extremely interesting premiums. One time it offered a kit of recipes for lemons and lemon juice along with a little plastic thing that makes juicing the lemon a snap. It's a nice idea and somebody, somewhere, had to invent that little plastic spigot. Look on the back of the next box of Kleenex you buy, and you'll probably find some kind of premium offer—anything from a potted plant (somebody had to invent the pot and mailer combination) to tiny needlepoint kits (somebody had to come up with the idea for the miniature needlepoint design). Cereal companies, soft-drink bottlers, canners, and even coffee makers are also good premium users. I once made a net $37,500 from an on-pack, self-liquidating premium offered by Quaker Oats on their puffed-rice and puffed-wheat packages. My product? Electro-static Stickies.

Basically, there are two different kinds of premiums. The first, and most common kind, is called "self-liquidating." With this kind, the customer who orders the special burger deal pays another two bits, lets say, and gets a hand puppet or a genuine Moldavian police whistle, or whatever the current premium is. The hamburger hustler bought the premium himself for two bits, so he breaks even on the deal. The premiums liquidate (pay for) themselves. The promoter can work the same trick with a "send one box top . . ." offer, except that the price of the premium also has to cover mailing expense and handling charge.

The second, less common kind of premium actually costs the promoter money. For instance, the same hand puppet or Moldavian police whistle might be given away free with the purchase of the jumbo-juicyburger combination plate. Usually this sort of premium is only used in case of a big event (maybe the opening of a new store), in the case of some special celebration (the puppets are handed out to kids from the hamburger hustler's float in the Christmas parade), or in the case of desperation (so far this month, nobody has bought the jumbo juicyburger combination plate and the regional manager's ulcer is starting to act up again).

In the case of premiums, the needs of the company and the mechanics of the operation are a little different from the usual retail setup, but you'll find that the ground rules for making the sale are about the same. When you make your pitch to the promoter, you'll have to do pretty much all the same things you would to pitch any retailer or a manufacturer's representative. You're especially going to have to prove your business stability and your ability to make enough of the product to meet the promoter's needs. Promoters simply cannot afford to have you punk out on them in mid-production. Hell hath no fury like a kid who didn't get his Moldavian police whistle.

Consider the two types of charts:

PREMIUM SALES

<table>
<tr><td align="center">(A)</td><td align="center">(B)</td></tr>
<tr><td align="center">CONSUMER</td><td align="center">CONSUMER</td></tr>
<tr><td align="center">⇧</td><td align="center">⇧</td></tr>
<tr><td align="center">PREMIUM BUYER</td><td align="center">PREMIUM BUYER</td></tr>
<tr><td align="center">⇧</td><td align="center">⇧</td></tr>
<tr><td align="center">PREMIUM REP</td><td align="center">MANUFACTURER</td></tr>
<tr><td align="center">⇧</td><td align="center"></td></tr>
<tr><td align="center">MANUFACTURER</td><td align="center"></td></tr>
</table>

(A) SELF-LIQUIDATING

The consumer pays the company involved	$ 1.00
The company pays the manufacturer	.90
The manufacturer's gross is	.90

(B) GIVE-AWAY

The consumer receives the item free.

The company involved pays the manufacturer, usually about	$ .90

When there's a premium rep involved in the sale, he'll normally receive about 10 percent of factory gross.

FULFILLMENT HOUSE

There's another way to turn your product into a premium. An even easier way. A "fulfillment house" might sound like some kind of group psychotherapy operation—or worse—but it's really a kind of distributor, specializing in premium items. Someone sends in a box top and an order, and the fulfillment house fulfills the order.

These firms have their own sales forces to do all the contract work with the premium users like soft-drink bottlers, cigarette companies, and so forth. They sell one of these users on your product as a premium, handle the shipping, maybe even the manufacturing, and do all the dirty work for you.

Your job is to sell the fulfillment house your idea. Then you can just lean back and collect your money—either a royalty or a flat fee for the idea.

The profit structure in this distribution system is about the same as in the distributor/wholesaler system I talked about earlier. Here's what happens to our hypothetical $2.00 item in this distribution system:

FULFILLMENT-HOUSE COUPON OFFERS

CONSUMER
(Through the company making the offer)
⇧
FULFILLMENT HOUSE
⇧
MANUFACTURER

The consumer sends in the coupon, boxtop,
or whatever (plus postage and handling) ....$ 2.00

The fulfillant house or offering company pays the
manufacturer anywhere from ....$ .10
                                                                      to
                                                                 $ .70

How come? Many of the coupon items you see offered are one-time purchases of factory closeouts. The prices get a little slippery. However, the reverse also happens. A manufacturer may use a coupon-offer sale as an introductory opportunity for a new product. The item, if it's successful in a coupon offer, will then be rolled out as a consumer item through normal distribution channels.

Keep this in mind if you can't really find any other way of underwriting your project. I personally have used this approach successfully twice. You'll be hard put to get any giant-number guarantees from a fulfillment house or any offering company. However, if it's a national program, you can expect at least 10,000 responses. From there, my experience indicates an average of 40,000 responses can be expected. And really successful offers can run into hundred of thousands.

INVENTIONCONVENTION.com

Another option of finding a licensing deal, investor, manufacturer, representative, or distributor is Stephen Paul Gnass' (he's the guy that did the Foreword) InventionConvention.com, an online networking cybershow on the World Wide Web.

Stephen has been producing the Invention Convention trade show for over fifteen years and told me that the online cybershow evolved as a natural offshoot of the tradeshow as a way to help bridge the communication gap between investors and entrepreneurs year-round.

Each invention is featured with its own Web page advertisement that includes a photo/drawing of the product, an extensive editorial-style product description, and an outline of the product's features and target markets. This cyberbooth (Web page advertisement) is online twenty-four/seven and can be seen by thousands of opportunity seekers every month.

However, Stephen also says that this shouldn't be the only thing you do to find what you're looking for. Stephen thinks you should be proactive, using this as one of many tools or vehicles in your marketing mix. Anyhow, you might check it out.

The final word on this chapter will encompass the infomercial, QVC-esque, and e-commerce opportunities that have become the apparent Distribution Darlings of the last decade. That word: *caution.*

10

Imagine, IF You Will...

A professional presentation can help sell your idea. Here's how to make one.

YOU ARE ABOUT TO DELVE INTO THE INNERMOST MYSTERIES OF THAT indispensable skill called "making the presentation." There's no getting round it. By the time you've had your run with your big idea, you're going to be a master of the arcane art of presenting your idea, in all its wonder and all its ramifications, to a lot of different kinds of people. It doesn't matter whether you end up selling your idea to a manufacturer, doing a small run of the product to generate some sales history, or forming a company of your own to make and sell the product. One way or another you're going to make a lot of presentations to:

* manufacturers

* bankers

* investors

* suppliers

* reps, distributors, retailers, consumers, patent attorneys, and, of course, your own family

The fine art of turning people on to your idea is going to become one of your most important skills. So it follows that you ought to have at least a basic understanding of how to prepare a presentation that will make the most of your bolt of lightning when it strikes. Maybe even help the lightning along a little.

FOUR SCORE and . . .

The first and most important thing to bear in mind is that your presentation ought to be as complete as you can possibly make it. It ought to answer all the questions anyone might ask—and some they haven't thought of.

Another primary consideration is the physical format. How it looks. What's in it. There are a couple of schools of thought on this one. For one, there's the old "back-of-the-envelope" approach. A couple of rough sketches, some written notes, all stuffed into a dime-store folder represents the farthest reaches of complexity in this approach. You rely totally on the power of the idea itself to blow people's minds, hoping that the humility of the approach will help things along.

Nick Underhill, when he was vice president and national sales manager at Entex Toys (then one of the giants of the toy model business) told me, "All I need to see is a rough sketch and a simple outline of the features. We know everything else about manufacturing and marketing an idea our way, so anything else would be extraneous."

On the other hand, another major toy company executive said, "We're all human and tend to react in our profession the same way we do in the rest of our lives. Therefore, the more complete and handsome the presentation, the more attention we seem to give it."

Which leads us into the alternate approach to the physical form of your presentation. Let's call it the "put-it-on-velvet" approach. Think of an $.89 ring. On cardboard, it looks overpriced. On velvet, it looks like a real value.

THE VELVET SCHOOL

Of the two options, I favor the velvet school. Regardless of what your idea is or to whom you're presenting it, if it's worth the time and money to make a presentation, it's worth the additional time and money to make as professional a presentation as you can. I know that the quality of the presentation made a sale for me many times over the years.

As an example, some time ago, I created a gift item for the man who had everything. It was a paperweight in the form of a giant headache tablet. But which one . . . Bayer, Anacin, Excedrin? At the time, Excedrin had the most ambitious advertising campaign, so it followed that theirs was the best coat-tail to grab. We wrote them a formal letter requesting permission to reproduce their E on the tablet and also use the name Excedrin in packaging and promotional material.

We received a prompt reply. "Absolutely not. Ours is a top-quality pharmaceutical product which cannot be associated with a piece of novelty junk."

This reply irked me. It was so arbitrary. We were not, in my opinion, in the junk business. Novel, yes. Junk, no. So I gathered together samples of all our existing products, along with reprints of our national publicity and our full-color catalogues, wrote another letter pointing out the quality of *our* products and put the whole enchilada in a box measuring about 18″ × 30″ × 6″. After wrapping it in white shipping paper, we covered the box with our vinyl flowers. The box fairly sparkled. A few days later I got a call from the Excedrin brand manager who said, "I had no idea what good stuff you're making and selling. Of course, we'd be happy to be associated with your new paperweight." I'm convinced the flower-covered box did the job.

What makes up a "velvet" presentation? A lot of simple things, beautifully presented. We've talked earlier about the value of getting good photography. You'll use it over and over again. One place is in your presentations. But before you take pictures, you have to have a subject to photograph. (Oh, the complexity of it all!) I always prefer photographs of prototype models to using drawings of an idea. It proves you've carried your homework one step farther.

If you do set up photography sessions for your product, remember to do one simple thing. Make that four simple things. Shoot each setup in:

* 4″ × 5″ color transparency

* 4″ × 5″ negative film (for color prints)

* 4″ × 5″ black and white

* Polaroid (for your own file reference)

If you do this all at once, it won't cost you much more than doing any one way by itself. And if your idea takes off, you'll save a lot of photo money later on. Incidentally, I favor 4″ × 5″ film. Your photographer may like 2.25″ × 2.25″ or 35mm. (If he *insists* on 8″ × 10″ film, get another photographer.) It won't ultimately make much difference, although some printers insist that bigger is better when it comes to reproduction quality. It's just that most of us take 4″ × 5″ a little more seriously. And this part of your job is very serious.

If you find yourself having to send several presentations to various groups at the same time, you should consider using slide copies of the originals. I've usually found them the simplest and least expensive ways to duplicate a presentation. Slides also command a certain amount of respect.

And, yes, I'm old fashioned. Digital photography—to my taste—is still too arbitrary and compromised to satisfy my antiquated demands. (I know I'm wrong, by the way.)

You might consider hiring a professional art director or an advanced advertising art student to help you develop your presentation if you choose the velvet approach. People in that line of work deal with presentations all the time.

COG/ASAP/FYI

So much for the graphic or picture side of your presentation. The words and numbers probably are more important. Especially the numbers. Exactly what you say in your presentation depends in part on whom you are trying to reach and what you want from them. But there are a few common elements to any presentation, whether you are using it to scare up some financial backing, trying to sell your idea to a manufacturer, wooing a sales rep, or whatever.

The first common element in sales performance are projections from a small market sample or from a research study you've done or simply guessed at on the basis of typical sales patterns for products of the type you're discussing. Obviously, the more solidly based your sales projections are, the more weight they are likely to carry. But one way or another, you'll want to show that your product has a glowing future in the market.

If you have real sales experience to back you up, spell it out in detail: how many units sold, how many retail outlets sold to, how many turns at retail, what time period is covered by the sales history.

Then, using that sample as the basis of your projection, state what sort of sales prediction you make for your product. No need to give a long-winded explanation of the logic you used to arrive at your sales projection. Give the person the background information and let him draw his own conclusions about the soundness of your thinking. (He will anyway, whether you like it or not.)

The second common element is your cost-of-goods calculations. Obviously, you've done these calculations fairly early in the game or you wouldn't be making presentations. All you have to do is spell out neatly and concisely all of the various costs involved in making and marketing your product, as we discussed in chapter 8. In addition to the obvious function of giving information on the cost of your product, this part of the presentation gives people an idea of just how realistic and professional you really are in dealing with the product.

Any patent applications, copyrights, etc., should also be noted and explained in your presentation. What did you apply for, when did you apply, what's the current status of your application? This gives people a feeling for what sort of protection your idea is likely to have and how professional you have been in your quest for glory.

Depending on whom you are presenting to, you may want to include cash-flow charts. If you are an accountant, you probably are familiar with cash-flow charts. If you are not, this book isn't long enough to explain what they are and how to make them. In general, a cash-flow chart is a graphic means of showing the income and outgo of money as you make and sell your product. Manufacturing costs money, selling makes money, but the crucial factor is this: *When* does the money come in and *when* does it go out? Ideally, there's always enough income to cover the outgo, of course, but at the beginning of a product's life that's rarely the case. You're always finding yourself one cycle behind on the cash-flow chart, needing to find funds to take care of the outgo long before it can be repaid by the inflow.

Other parts of the presentation will be dictated by the audience you are addressing. But in all cases, the objective to keep in mind is to present yourself as a professional, realistic businessman and to present your product as a potentially successful and profitable one.

GREAT PRESENTATIONS SELL

One good place to look early in the process of building your presentation is a bank. Get together with one of the loan officers and find out the kinds of information he would want to see in a presentation. You may not be interested in making your presentation to a bank at all, but the kinds of questions a banker has about your product will be indicative of the kinds of questions other people will be asking. If you can put together a presentation that satisfies a loan officer at a bank, you can probably satisfy anybody. Think of it as a dry run. (In my experience, that's how conversations with bankers usually turn out anyway.)

An accountant is another excellent source of information on what to include in your presentation. In fact, it would be well worth your while to hire an accountant to help you put the presentation together, especially if you are a little shaky on topics like cash flow.

Frankly, the whole topic of presentations could fill an entire book by itself. I don't claim that this short treatment is exhaustive. All I'm interested in doing is convincing you that a professional presentation, even for the simplest of ideas, is important. If you can do it yourself, fine. If you have to hire professionals to help you, do it. One way or another, get a good, solid, well-turned-out presentation because it can mean the difference between success and failure. If I had to bet on a terrific idea with a lousy presentation versus an average idea with a terrific presentation, I'd bet on the great presentation every time.

I THINK I'M the SCREWEE

There's one last subject to be covered under this general heading of presentations. It's the "corporation idea submission form." This is the old light-under-a-bushel problem that ultimately confronts every inventor who wants to sell his idea but is afraid of having it stolen. And that's *everyone* on his first time out. You *and* me.

Applying for a patent ought to alleviate some of your fears but not all of them. And what if you haven't applied for a patent, or you can't get one? You could become the screwee, or at least you could worry about it. (If you are presenting your idea to a well-established company, the odds against that danger are really quite long. But that doesn't stop those nightmares.)

One option that my friend Stephen Gnass, executive director of The National Congress of Inventor Organizations, recommends is filing a Provisional Patent Application (PPA) with the U.S. Patent Office. It costs only $75 to file, and once the Patent Office receives it, the inventor can legally say that he/she is patent pending. Stephen adds, "The GATT Treaty of 1995 enabled inventors to apply for a PPA prior to having to apply for the Regular Patent Application which will cost thousands of dollars."

"I recommend a PPA in that it gives inventors up to a year to find a licensee or buyer for their idea without enormous expenses during this phase. Yet they can say they're patent pending, which will help when they are making presentations." Since it's a relatively new approach, and I'm from the old school, I, myself, am not as sold on the PPA approach yet. However, it's important that we include this important new option.

Now, back to the nightmares. The submission form many companies ask you to sign isn't going to help you feel any better. Alcoa has permitted me to quote from their form here. It will give you an idea of the sort of thing you'll encounter. I like the way Alcoa goes about it. Their forms are just about the same as any you'll ever see for idea submissions. But Alcoa goes to the trouble to accompany the form with a thoughtful booklet explaining their position on new ideas from inventors. Still, what the form says is hardly reassuring:

———, 20——

New Ideas Section
Aluminum Company of America
1501 Alcoa Building
Pittsburgh 19, Pennsylvania

Gentlemen:
I have an idea relating to ———————————————————————
which is fully described in the attached material or in material which I have already sent you. Although you have not asked me to submit this idea, I would like to have you examine it to determine whether it may be of some value to you.

I realize that, to avoid misunderstandings, ideas must be submitted under certain terms and conditions before they will be considered. Accordingly, I am submitting this idea to you on the following terms and conditions which shall apply to this idea, including all related material, and any supplemental ideas and material which I may subsequently submit.

(1) I am not submitting this idea in confidence. I agree that my submission of this idea and your consideration of it shall not create a confidential relationship between us, expressed or implied, and that you have no obligation to keep this idea a secret.

(2) I understand that you will consider this idea only to the extent that you believe it is deserving of consideration; if you decide that you are interested in the idea, I would appreciate you informing me of that fact.

(3) If you choose to do so, you may enter into a formal written agreement with me relating to this idea which may, among other terms, provide for the payment of compensation to me. However, I agree that you will be under no obligation to me with respect to this idea unless and until we enter into such written agreement, and then only

in accordance with the terms of that agreement. I further agree that neither your consideration of this idea nor any discussion we may have concerning it shall be construed in any way prejudicial to you.

(4) If you decide that you are not interested in this idea, I understand that you will inform me of that fact, but I agree that you will have no obligation to advise me of the reasons for your lack of interest, nor will you be obligated to return any of the material which I have submitted, reveal any of your activities in connection with your consideration of this idea, or to reveal any information pertaining to your research or other activities in any field of technology or business relating to this idea.

(5) I agree that the consideration you give to this idea shall not constitute a waiver of your right to contest the validity of any patent which has been or may hereafter be issued on it. Should I contend that you are infringing on such a patent, my sole remedy will be that available to me under patent laws.

(6) I agree that these terms and conditions cannot be modified or waived except in writing signed by an officer of Aluminum Company of America.

Very truly yours,

The first time you read one of these "Corporation Submission Forms," you'll probably react as I did. You'll shout and scream a lot. Things like, "Rape!" But think about it a bit, and you'll realize that your idea is probably one of thousands presented in any given year. And whatever the company, they are developing ideas every day of the year themselves. They, too, must be protected from the proved psychological phenomenon called "Concurrent Development." Ideas somehow find their moment in history. All over the country, in fact, all over the world, several people can be working on the same idea simultaneously and never know it. So the form you are asked to sign protects the company from your later coming back and claiming they stole your idea. Believe me, their records on ideas are a lot more complete than yours will ever be. So what to do? All I can tell you for sure is what I do.

I sign on the dotted line.

There is one gambit you can counter with, however. It's called an "Agreement to Review Idea" form. Here's an example:

AGREEMENT TO REVIEW IDEA

We, the undersigned, agree to receive in confidence full details about an idea for a new product to be submitted for our consideration by Billy Smith.

It is further understood that we assume no responsibility whatever with respect to features that can be demonstrated to be already known to us. We also agree not to divulge any details of the idea submitted without permission of Billy Smith or to make use of any feature or information of which the said Billy Smith is the originator, without payment of compensation to be fixed by negotiation with the said Billy Smith or his lawful representative.

It is specifically understood that, in receiving the idea of Billy Smith, the idea is being received and will be reviewed in confidence and that, within a period of 30 days, we will report to Billy Smith the results of our findings and will advise whether or not we are interested in negotiating for the purchase of the right to use said idea.

Company _____

Street & Number _____

City _____ *State* _____ *Zip Code* _____

Official to receive disclosures (please type)

_____ *Title* _____

Date _____ *Signature* _____

Accepted _____

Billy Smith, Inventor

The purpose of sending the company this form, instead of the one they'd like to have you sign, is this: Your submission form gives you a little more protection and restricts the company a little more than its form. You'll notice that this one requires the company to keep your idea confidential, puts a time limit on the company to review your idea, requires the company to report to you in writing on the outcome of its review, and even requires the company to pay you for your idea.

From the inventor's standpoint, that's wonderful. But, frankly, I don't think much of these forms. Maybe I'm a little old-fashioned, but I still feel that the buyer has a stronger hand than the seller. If somebody wants to sell an idea, he ought to be the one who does the signing of forms. He needs the buyer much more than the buyer needs him.

PLEASE BUY my IDEA, you CHEAT

The "Agreement To Review Idea" form I've reproduced here resulted in a perfect example of the reaction these forms usually get. I signed this one recently, mostly to get a current example of this kind of form. Billy Smith, the inventor, came to me with an idea and wouldn't reveal it to me until I signed the form. I signed. And sighed.

The idea turned out to be another spin-off of an old idea. I wasn't interested, so I gave Billy the name of a company in Chicago I thought might be interested.

Billy sent his idea to Chicago, along with one of his idea review forms. I told him I didn't think that was a very good idea, since the company was certainly reputable and might not take kindly to that kind of treatment. But Billy insisted, and the day after the form arrived in Chicago, the man in charge of new ideas at the company called back. He had a brief, succinct, gravely voiced suggestion on how Billy might best dispose of his idea. And his tightly rolled idea submission form.

A Miracle, JUST A Little One

Turning a major marketer's interest into a lot of free help. It's called leverage.

SOMETIMES IN THIS STRANGE AND WONDERFUL BUSINESS OF CHASING rainbows, it seems that the only thing that will help you survive is a miracle. Just a little one, but a miracle nonetheless. Okay. Here's a small miracle for you. Here is how to find a big, influential helper who can:

* Put you in touch with the right manufacturers
* Help you find materials to make your product at a good price
* Get technical help for you
* Carry enough weight in business to open some sticky doors
* Give you a hand in hacking your way through the brambles of business
* Do it all for free

The key to this miracle is simple. It's called profit motive. If your idea looks as if it could make a lot of money for some major marketer, he'll be your friend. If a buyer for a major national retailer or somebody of that ilk thinks your product can be a hot seller in his stores or catalogs, he's going to be on your side. He's going to *want* you to succeed so that he can make a lot of money selling your product. In order to help you get your idea off the ground and onto his shelves, he'll do whatever he can to smooth your path. It's simply good business on his part.

The amount of help you get, of course, will be proportional to the amount of excitement your idea generates. Regardless of the degree of help you get, however, there's nothing unusual about it. I call it "leverage," and I run into it all the time. I'll give you an example of how it worked for an inventor just like you.

Incidentally, I was going to update this example just so it looked like I was paying attention. However, after rereading this twice, I've concluded the content is timeless and the points could not be made any better with more current examples. Leverage, it turns out, is timeless. And priceless.

MARKETING GARBAGE

The subject of our example is garbage. Specifically, a new way of dealing with garbage. That may not be a big turn-on for you, but garbage is important stuff these days. As we become increasingly aware that we run the risk of trashing ourselves off the face of the earth, we become more and more interested in garbage. It has become one of the big issues of our time.

For years, garbage has been just sort of indiscriminately tossed out with no real effort to sort it. On the whole, garbage has been treated like—well, like garbage. But today, there's a whole industry springing up around the job of reclaiming paper, glass, and metal and putting them back to work.

Most communities around the country now have laws requiring that trash be sorted, separating the various reclaimables from the coffee grounds and egg shells before they get the old heave-ho. Many more communities are passing such laws every year. Even where it isn't required by law, some conscientious people are sorting the trash anyway. They're doing their civic duty—and making money, too—by sorting out the cans, bottles, and paper and selling them to recycling centers.

But sorting out garbage is a drag. Either you have a whole bunch of separate containers cluttering up the kitchen or you have to go digging around in the garbage after the can is full. Yukkkhhh! There has to be a better way, right?

Well, a Sears buyer friend of mine certainly thought so. As the ecology thing began developing into an important factor in our lives, my friend was already looking at it from the marketing standpoint. A trash sorter was one of the things that sprang to his mind as a potential product. He asked the Sears research staff to run down some numbers for him on how many people out there were separating their garbage into the various reclaimable categories. The answer was mind-boggling.

Fully 15 percent of the people who match the typical Sears customer profile are sorting their garbage! That worked out to something like 12 million customers and a $150 million market for a neat, easy-to-use, inexpensive gadget to sort trash, according to their projections.

Frankly, that sounded kind of high to me. But my friend seemed to know what he was talking about when it came to those things, so I took his word for it. Either way, there was obviously one huge market out there for a trash sorter.

Needless to say, my friend had very little trouble cranking up a design project to develop a sorter. For three years, the Sears heavy-thinking-and-design types wrestled with the problem and got practically nowhere. Nothing. Then, in the middle of that frustration, my friend got a phone call that changed everything.

The caller was a bright, attractive woman named Suellen McDonough. She was a mother of four, a part-time nurse, and the wife of a manufacturer's rep in Durham, New Hampshire. Mrs. McDonough had developed an invention she hoped my friend might be interested in.

"What is it?" he asked.

"Oh, it's sort of a trash separator," Mrs. McDonough replied. "You know, for sorting out bottles and cans and stuff."

A few minutes later she had spelled it all out. It was so simple. The whole thing, which she called a "Recycl-it Basket" is nothing more elaborate than a three-hole wastebasket. It had a one-piece plastic molding with three big pockets in a row, just the right size for the standard grocery bag. (It was ecologically sound in that respect, too, since most grocery bags were made of recycled paper.) It doesn't take up much more room than a regular ol' one-holer and it made the whole process of sorting trash so simple you could do it in your sleep. No big technological breakthrough, no transistorized frammistats and microcircuitry, no devilishly complicated machinery. A plain old three-hole wastebasket. And nobody had thought of it before.

Here was an ideal example of the little miracle I promised you. We had a major marketer interested in an inventor's product. And we had an inventor, Mrs. McDonough, who knew exactly what she was doing. She had covered all the right bases, done all of her homework along the way—just as I keep nagging at you to do yours. There were no dangling loose ends to disconcert anyone. It's worth stopping for a moment just to see how she went about it, because she's a perfect paradigm of how to give your product its best shot at the marketplace.

THE LITTLE ENGINE that COULD

Back when her invention was still an idea on a piece of paper, Mrs. McDonough went to the local office of the New England Industrial Development Resources Center where she got first-class advice and counsel. (There are a number of similar governmental and quasi-governmental organizations around the country.) The center put Mrs. McDonough in touch with an industrial engineer who explained that there were two radically different ways to go about manufacturing her basket.

On one hand, she could spend the relatively modest sum of $2,000 or $3,000 to get a set of molds for a process called rotocasting. Rotocasting, he explained, is fine for small production runs. The only drawback is that it is a slow way to make plastic objects, which means that the price per unit is quite high.

On the other hand, the engineer continued, she could invest somewhere in the neighborhood of $40,000 to $50,000 for a set of high-speed injection molds, just like the big kids use.

That didn't sound like her kind of neighborhood, so Mrs. McDonough had a set of the cheaper rotocasting molds made. Then she ran off a sample batch of baskets and took them to Jordan-Marsh department stores in Boston where she made a sale. Shortly, it became apparent that the basket was creating a stir in her rather limited market, but the rotocasting process was totally unsatisfactory. It wasn't just the cost. The process simply didn't do a good job of making her rather large baskets.

Still, how could she afford to tool up for a more sophisticated manufacturing process? She decided to reach out for more stores and wider distribution so she could make the heavy investment in molds feasible. She went to New York to talk to buyers and manufacturer's representatives. Nothing much came of those visits, but the trip to New York wasn't a total loss. As long as she was in the Big Apple, she decided she might as well have a go at Sears.

There was a moment of chagrin when she discovered that Sears doesn't believe in the Big Apple. Their headquarters are out in the Midwest, like God intended. Nothing daunted, she dialed long-distance information in Chicago, got in touch with the Sears switchboard, and asked who bought trash containers. She was connected with my buyer friend who, up to that moment, didn't know her from Adam's house cat.

You might say it was all a lucky fluke. But as a matter of fact, there wasn't any luck involved. Mrs. McDonough was taking all the correct steps, and they led her unerringly to the man who wanted exactly what she had to sell.

If you do your homework—the golden rule of the idea business—you can be every bit as "lucky" as Mrs. McDonough.

HOW about $50,000 'til PAYDAY?

Here's where we get into how the buyer helped her. Sears, like many major marketers, was not a manufacturer. They don't make many of the things they sell. Instead, they contract with manufacturers around the country—big ones, little ones, middle-sized ones—to make the products for Sears. So in order to sell to Sears, Mrs. McDonough would either have to become a manufacturer herself on a large scale, or she would have to make a contract with an existing manufacturer to make the baskets for her. Obviously, her little rotocasting setup wasn't going to be up to the job of filling the orders the buyer was thinking of. Equally obvious, she didn't have $40,000 or $50,000 lying around in the sugar bowl at home to buy high-speed injection molds.

So the buyer gave her the names of three plastics manufacturers in her part of the country, all of whom had done business with Sears before and all of whom were capable of doing the job for Mrs. McDonough. He told Mrs. McDonough to let them know that he was definitely interested in her product and that he would be happy to confirm that for them.

Now *that's* a valuable piece of leverage! Anytime you can walk in and tell some manufacturer that Sears is waiting in the wings with an order book, you are going to have a whole lot of undivided attention. Notice that there was no concrete order at this time. Just an indication of interest. That's all you can usually expect, and that's all it takes, most of the time.

It's no guarantee of success, however. In the case of Mrs. McDonough, two of the three manufacturers said they just weren't in a position to take on the project at the moment. The third was interested but wanted Mrs. McDonough to pay for the high-speed injection molds herself, then let them manufacturer the separators and pay her a royalty.

That was a bad deal. After she'd borrowed the money to pay for the molds, she'd still be on the outside looking in. All she would get is a royalty, but no control over the operation. She'd have to sell an awful lot of trash separators just to break even. (You can argue that the manufacturer has the same problem, and you'd be right. But amortization—spreading the cost of the tooling over a predetermined number of products in the production run—is part of the manufacturing business. It's one of the costs the manufacturer would normally calculate in determining its price. Mrs. McDonough shouldn't have had to pay that cost for them.)

The buyer felt Suellen could do better than that, although Mrs. McDonough was ready to go out and borrow the money to pay for the molds. He was convinced that they could find a manufacturer who would invest in the molds himself, then make the separators under contract to Mrs. McDonough.

It took quite a bit longer than anyone anticipated, but eventually the buyer found the right manufacturer. A company in North Carolina has agreed to make the baskets for Mrs. McDonough under a royalty arrangement.

NO GUARANTEE of SECURITY

As you can see from the example of the trash sorter, a big marketer is in a position to do you a great deal of good if the marketer is considering carrying your product. In the first place, any manufacturer has to be interested in making your product for you if he knows that a major outfit is willing to carry it. It indicates to the manufacturer that there's an existing market for the product. It takes a lot of the risk potential out of his tooling up for the project.

Second, it indicates to the manufacturer that the inventor is not just some kook with a hare-brained scheme. If you have stirred up some interest at a major national retailer, you might be onto something pretty substantial. Big outfits don't fool around with weirdos and dingbats.

Also, with an expression of interest from a big marketer, financial backing for your idea becomes much easier to find. Few bankers or investors feel very secure about their personal understanding of the retail marketplace. An expression of interest from a major marketer gives them a lot more confidence about your ideas. It takes some of the guesswork out of it for them.

Finally, with a handle on a big marketer, you can make better buys on materials for your product, and you can get a better price of manufacturing. The bigger the order, the lower the price per unit, usually. And when you're selling to a J.C. Penney or a Sears, you're talking in big, round numbers. With a lower cost of goods, you can make more profit, and you can sell at a more attractive price, which in turn ought to increase your sales.

You may be thinking that the marketer who helps you could just as easily knock you off. There is *no* guarantee of security anywhere in this book, nor anywhere in my experience. You always run the risk of having your idea stolen. The only real protection you can give yourself, at this point in development is—you guessed it—careful, complete homework. Here's what I mean:

There you are, showing your product to the buyer for some major marketer, and he's interested. Then he looks at your cost of manufacture and knows he can beat that price substantially. If he's truly a fine person, he'll point that out to you and probably help you find ways to bring down the cost. But there *are* people out there who are not so fine. They may simply turn you down, then get on the phone to a couple of suppliers and be out on the market with the same product—your product—only cheaper.

If you had done your homework, you would already have found out how to make it cheaper. If your price is pretty close to the best price he could get, he'd just as soon buy the product from you. It's a lot simpler that way.

So do your homework thoroughly and protect yourself.

THAT'S ONE NEAT GARAGE

There are times when the kind of leverage I'm talking about here could make the difference between getting on the market at all and simply packing it in as a lost cause. My case in point is an invention I developed some years ago. It's a modular, prefabricated system of interlocking storage cabinets for home garages. I call it Neat Garage.

Just a bit of background on the product, before we get into the details of how I went about trying to get it sold. All you have to do to establish the need for Neat Garage™ is to take a look at your own garage. If it's like most, there are rakes and shovels and garden hoses and bikes and woodworking tools and a Christmas tree stand and Lord knows what all scattered all over the place. You're lucky if there's room for a car in there at all, among all the accumulated oddments.

What to do about it? Well, you could call in a carpenter at thirty or forty dollars an hour and have him build you some storage cabinets and lockers around the garage. Or you could have him build you a storage room as an addition to your garage. You might even put up one of those corrugated metal storage sheds in the back yard—the sort of prefab building you can buy at hardware departments.

Or, you could use Neat Garage. The cabinets hang on the walls, interlocking with each other, taking up very little of your garage space. With built-in shelves and hanging hooks, they allow you to organize that incredible mess of junk and tools so that you can find it when you need it and keep it out of the way when you don't. Being prefabricated, the storage is a good deal less expensive than custom-made storage units. And putting it up is so simple, you can do it yourself.

It's a natural for some big cabinetmaking company or some big plastic molding company or some big aluminum company. It is definitely *not* a natural for someone like me to manufacture and market on my own. Tooling up for this one would cost something like the Bolivian national debt. I was not about to spend that kind of money, even on the apparently nonsensical assumption that I *had* that kind of money to spend.

Now, I had done my homework on this one. I'd made some pretty realistic market projections, based on the number of new homes being built. (A natural market for Neat Garage is the company that builds those new homes. Another natural market is the guy who buys a new home and instantly discovers he has no place to store all that junk he's accumulated.) I worked out my cost of manufacture as closely as I could. I worked out all the bugs in the design. I even built full-scale models (which explains why my garage doesn't look as messy as the average garage anymore). And I put together a very uptown presentation portfolio, complete with $8'' \times 10''$ before-and-after color glossies, sales forecasts, statistics, and such. I was loaded for bear.

But for quite a while I couldn't find the right bear. Boise-Cascade took an option on it for two months and did its own market research. It assigned a team of experts and spent weeks coming to all the very same conclusions I had come to. Yes, there was a market. Yes, they would sell a whole bunch of Neat Garage units. No, they weren't interested. At that time, Boise-Cascade was looking for new products that would give a 50-percent return on investment, and it figured Neat Garage would return only 30 percent. Now, in *my* book a 30-percent return wasn't all bad. But the company was shooting for the big stuff, so it turned my idea down.

Another company I presented to just wasn't turned on by the idea. As simple as that. So there I was with what is unquestionably a solid, marketable idea. Only I couldn't get any manufacturer interested in buying it from me. And I was in no position to market it myself.

As you will discover later on in the book, a manufacturer has turned up with an interest in my idea. It's beginning to look as if I might sell Neat Garage.

But suppose I'd never met that particular manufacturer? Or suppose, after pondering the idea, he ends up turning me down? Then there would be nothing left but Plan B. And it's not a bad plan.

Plan B is leverage. If I can't find a manufacturer who'll pay me a royalty on my idea, then I'm going to go to a major national retailer—just like Mrs. McDonough did with her trash separator. I'm going to show them my first-class, uptown presentation portfolio and explain how the price structure would work out, how many sales I predict, and so forth. And I'm going to ask the retailer if it'd be interested in carrying it if I could find a manufacturer.

My guess is that the retailer will say yes, based on the price and the specifications I will show them. Then, with that expression of interest in hand, I'm going to go back knocking on the doors of a lot of manufacturers. I'm going to use leverage—to get a manufacturing deal. My objective, of course, is to sell the idea for a royalty and bring the manufacturer and the marketer together, introduce them, and tiptoe out of the picture.

In fact, if my idea really lights that national retailer up, they might be able to help me find a manufacturer and do everything but hold my hand while I negotiate the deal.

There are a million variations on how you can use the leverage gained from interesting a big marketer in your idea. With somebody like a major buyer to help you in areas where your own expertise may be sketchy, you'll find the going a lot easier. And with an expression of interest from one of the biggies in your back pocket, you'll find a lot of doors opening for you in your quest for a manufacturer. Just how you use the leverage depends to a large extent on what your idea is.

Obviously, major buyers in the business aren't around for the sole purpose of helping every inventor who comes down the pike. But they are all interested in developing new ideas that fit their markets. If you can turn one of them on with your product, you've got more than a big sale.

You've got an important ally on your side.

Let's Make a Deal

How much to ask for, how much to expect when you sell your idea.

OVER AND OVER AGAIN, I'VE BEEN TELLING YOU THAT THE ONLY WAY you really want to go with your invention is to sell the idea to a manufacturer on an advance-against-royalty deal. Now it's time to talk about making that deal. It's one place where many inventors, who have been shrewd, calculating, and industrious all along, suddenly become bumbling amateurs. And the single factor that trips up most of them at this point is plain old greed.

GREED SUCKS

Don't let greed get in your way when you're negotiating the sale of your product to some manufacturer. Most inventors, unfortunately, have a grossly inflated notion of the value of their idea and their rightful share of the product's sales price. In going into their negotiations with that attitude, a lot of inventors blow their chances of getting something going. They make excessive demands relative to the value of the idea and wind up with 100 percent of nothing.

Don't forget that it's the manufacturer who's going to be taking all the risk with your idea. My friend at Sears put it this way: "Take for example, the decision to spend $60,000 on a mold. The product could fail for any of a hundred reasons. It's the manufacturer who usually must make the hard-dollar investment. *Investment* is the key to a new product, not the idea. It's also a matter of priorities. With the inventor, his idea is *always* the number-one priority and obviously (to him) very valuable. To the manufacturer who is about to gamble maybe $100,000, that idea could fall well down on his list of ways to lose money."

If the profit on your product is big enough to allow it, a legitimate company will be happy to pay you a reasonable amount of royalty. But if your demands cut into his reasonable and proper profit on manufacturing and distributing the product and risking his capital, he'll balk. And he'll be right. Your excessive demands can only show up in one of two ways: increased price of product, which could price it right out of the market, or in a lower profit for the manufacturer, which could make your invention very unattractive to him. You have to be realistic.

If you've done your homework, you ought to know how much profit spread is available to play with in your negotiations. You can be sure that the company you're negotiating with will know where the pennies are buried. Personally, I'd guess that there is rarely more than a percentage point worth of real negotiating room to play with. An offer of 4 percent, countered by a demand of 6 percent is about as freewheeling as these things get. Over the years, I've come to the conclusion that the average royalty for a reasonably well-protected, well-developed product is about 5 percent of factory gross sales.

FIVE PERCENT is ABOUT it

While we're on the subject, let's clarify that term "factory gross." Although there may be a slight variation of interpretation from company to company, the following is generally what you should have in your mind when negotiating a deal that will be fair to you and to a manufacturer.

You arrive at factory gross by starting with the price the factory sells an item for. Let's say, for this example, it's $1.00. For that selling price, the cost of sale is subtracted. As you've learned, that commission can range anywhere from 5 percent in the toy industry to 20 percent in the gift and stationery field. We'll use 10 percent for this example. Ten percent of $1.00 is $.10 (as I said early on in this book, you don't have to be a genius to succeed), so you're now at $.90. From that you subtract discounts for other elements of the sale such as cash discounts (usually around 2 percent) and freight allowances (usually around 3 percent) or a total of 5 percent. *Voilá!* You get $.85 as the factory gross. That's the base number for figuring your per item royalty. Using a 5 percent royalty, do a little multiplication, and if you come up with $.0425, we agree that is your royalty per item sold. If the factory sells only one, you won't be retiring any sooner than you had planned. On the other hand, what if they were to sell 10 million? . . .

In the process of writing this book, I talked to a lot of people in the business, and without exception they all gave me 5 percent of factory gross sales as their estimate of the average royalty deal. I didn't even have to prompt them. In a high-risk industry or where the product is not as well protected from knock-offs, a royalty as low as 2 percent might be eminently fair.

Here's yet another example: A friend of mine brought me an idea he had spun off of Rickie Tickie Stickies. But he'd carried it into a whole new direction and had what seemed to both of us to be a highly saleable product. He had taken some of my flower designs, and some other designs of his own, and translated them into heat-bonding patches. Bill called them "Snappy Patches." You could just take one of them, lay it on a pair of Levi's or a jacket or something, make a couple of fast passes with an iron and you'd have a funky design permanently affixed to it. And if you patched a rip, all the better! This, mind you, was more than a year before the whole patches craze swept the country. Bill was on top of a great idea well ahead of the pack. But we couldn't make a deal.

Bill wanted me to make and distribute the patches and pay him a royalty. Fine. I had the venture capital available, and I thought the idea was worth it. The trouble was Bill wanted 30 percent of gross as his royalty. I laid out all the figures—cost of manufacture, all the other costs we went over a few chapters ago—and showed him in black and white that if I paid him 30 percent, he was going to be making something like three times what I was making. And I'd be taking all the risk. All he had to do was sit back and collect his royalties. Bill didn't see it that way at all. I offered him 5 percent, which was a pretty generous offer, under the circumstances. No sale. He was determined to take it someplace else and get his 30 percent. Or, failing in that, he'd do it himself. I shook my head sadly and wished him well.

A couple of years went by and once again Bill was in my office. He'd learned his lesson. He'd tried and failed to get his Snappy Patches off the ground on his own. Now he was ready to discuss that 5 percent royalty deal. But it was too late. The fad for patches had come and gone and the few patches still around were being discounted just to get them off the shelves. Bill, who had been sitting on top of a dynamite idea a year before the craze hit, never made a dime from it. He—like yours truly before him—had fallen victim to greed.

If the money had been there, believe me I would have been happy to give him his share. Later on, another fellow came to see me with a great idea called "Parent Protest Posters™." They were take-offs on some protest rhetoric of our era, neatly reversed so it was the older generation doing the protesting. For instance, an old lady in duster cap and apron looking exactly like Flagg's Uncle Sam in drag was saying, "Mom wants you to clean up your room." They were to the point and well done. My friend wanted me to manufacture and distribute them, printing the posters from his designs. They're still selling, although the bloom is pretty much off the rose by now. Now, this was an entirely different-colored horse from Snappy Patches. The cost of goods was low, so there was plenty of elbowroom for a fat royalty. We agreed on 10 percent, mostly because it would have been embarrassing to offer him anything less. The posters sold very nicely and we both made money. I was happy, under those circumstances, to give a bigger-than-normal royalty deal.

ONE HUNDRED TWENTY PERCENT of GROSS

In the process of researching this book, I did a little experiment of my own, just to see if what I'm talking about is really true. Way back when I got stung on the Little Lumpsie doll and finally sold it off to Fisher-Price, I was glad to get a 2 percent royalty. At the time, the negotiation on that price took about thirty seconds. Bob Hicks asked me what I had in mind. I said I was thinking about 120 percent of gross. What was he thinking of? He grinned and replied, "One percent." I asked if he'd settle for 2 percent, and the deal was done. Some years later, I was picking Bob's brain for the original version of this book, and I reminded him of the doll. "Do you suppose you would have gone to five percent at that time?" I asked.

"Nope," he said firmly. "There were too many unknowns about your product, even though you had some good sales results to show us."

"How about four percent?" I asked.

"Nope."

"Three?"

There was a pause. He smiled.

"Yes."

So there it was. We had about a percentage point to play with in our original negotiations. I took a lower figure than, as it turned out, I might have been able to get. But at the same time, I got an advance of $5,000 as part of the deal, and I was happy to get out of the doll-making business.

OPTION, SCHMOPTION

Speaking of up-front money, there's another area where many inventors have unrealistic ideas. I know I did. The area is that of options. Boise-Cascade took an option on my Neat Garage idea, as I mentioned earlier. The deal was that for two months I wouldn't sell it to anyone else while they did their homework on the idea. I was delighted. Visions of five-figure options checks danced in my head. I asked how much they were going to pay for the option, and they looked blank. "Five thousand, maybe?" I asked, beginning to suspect that this wasn't a five-figure deal after all. Again, no reaction. "Well how about five hundred?" Pause. "Hey, look, I paid $240 to fly up here, could I maybe get that back?" No deal. They explained that it is not their policy to pay for options. Later, if they took on the idea, they might consider an advance against royalties, but that's about as fast and loose as they ever play with the company checkbook.

In a way, maybe I'm being a little unfair to myself and my fellow inventors. Often as not, it isn't entirely a question of greed, but of simply not understanding how the business of marketing a product works. A while ago, I was playing matchmaker between an inventor and a rep in a deal where a few weeks earlier there had been real animosity in the air. All because of failure in communications between two honest, well-intentioned people over a rather interesting new product.

HE SAID, SHE SAID

The inventor was an attractive housewife who lived in Palos Verdes, California, the peninsula where I had my home and my offices. Sondra Cutliffe was into arts and crafts, and she had a highly imaginative mind. She turned those assets into a fascinating hobby for some years. She liked to design Christmas decorations, often making them out of common household objects. The year in question, she had designed a clever little doll decoration made from a clothespin, some baker's clay (a dough which turns rock hard when it dries) and a little paint and glitter. The result, in her hands, was a delightfully, old-fashioned-looking Christmas tree ornament.

In the past, Sondra had made a few dozen of the Christmas decorations each year and sold them to a few gift stores, making a modest profit and playing at business. This year, she decided to up the ante a little. She'd been successful enough on her own, just taking the ornaments to local stores. What would happen, she wondered, if she took the idea to a manufacturers' rep? Palos Verdes doesn't exactly rank as a poverty pocket, so there would be enough housewives around with time on their hands to form a cottage industry at Sondra's house if the demand got high enough. For the fun of it, of course.

Doing a little checking around, Sondra finally decided on a bright, aggressive, then young, manufacturer's rep named Roger Gruen. Roger sold Christmas stuff—tree lights, ornaments, tinsel, and Christmas novelties—so his line was an ideal place for Sondra's product. She showed Roger her clothespin ornaments and he liked them. She told him she could probably make as many as 8,000 of them if she got the neighbors together to form a kitchen-table assembly line. That hardly qualified her ornaments as one of Roger's hot items, of course, but he liked the pieces enough to add them to his line for the upcoming trade show in Los Angeles. That's where things got exciting.

All the retailers in the area who sold Christmas items came to the show to see what was new in Roger's line for that year and to place their orders for the holiday season. One of the things that really turned them on was—you guessed it—Sondra's ornaments. The day before the show officially opened, while everybody was milling around trying to set up booths and exhibits, a few buyers managed to sneak in. Those few buyers placed orders for nearly eight hundred of Sondra's ornaments. And that was before the show even opened! Three days into the show, Roger had orders for 13,000 ornaments, and he knew he was on to something.

In the midst of the hullabaloo of a major trade show, he did some fast calculating. His orders had outstripped Sondra's ability to manufacture the ornaments. He quickly made a decision to manufacture and distribute the ornaments himself and work out a royalty deal with Sondra. He figured he could find places to have the ornaments manufactured in the quantities he needed to fill his orders. With her design talent, plus his investment, his contacts, and his sources, they could have a really hot product on their hands.

In the meantime, Sondra dropped in at the show, saw 13,000 orders and very nearly fainted. "But, I can't *make* that many!" she wailed.

"Don't worry about it. I'll have them made," Roger told her, flushed with success. In the madhouse of the show, surrounded by surging crowds of buyers, he tried to explain to her about royalties. But all Sondra got out of it was that she was going to be making a whole lot less per ornament than she had agreed to originally. And how on earth was she going to make 13,000 ornaments anyway? She was highly upset and Roger, sensing he was about to get into a real jam, took the ornaments out of his display to forestall disaster. He didn't show them anymore for the rest of the trade show. That was no time for him to conduct a primer course in marketing.

Sondra was convinced that she was being ripped off by Roger, so she called a lawyer and had him call Roger to demand that he cease selling the ornaments and turn all the orders over to Sondra. Meanwhile, she asked around the neighborhood and found out that there was this Kracke fellow a few blocks away who was supposed to know something about inventions. She called me and asked to come talk with me, which she did. And I found myself in the middle of no man's land with a shooting war about to break out between Sondra and Roger.

As far as Sondra was concerned, Roger was stealing her idea and going into business on his own, paying her a mere pittance on each ornament, instead of the whole enchilada minus a sales commission. He had even covered over her name on the display at the show and replaced it with his own name. What's more, he had taken orders for an enormous number of ornaments, knowing full well that she couldn't produce that many.

For his part, Roger felt he was being unjustly attacked by an artsy-craftsy suburban housewife who couldn't or wouldn't understand that she had a potential success on her hands and needed his help desperately. He had the means at hand to make them both a lot of money. But as manufacturer and distributor of the ornaments, he was entitled to a fair profit, too. He couldn't very well take all the risks and do all the work for a 20-percent sales commission—if only she could understand that (a) she *didn't* have to make the ornaments herself, (b) he was willing to risk his own money on her idea, and (c) he was being totally fair and aboveboard in the whole thing.

I have to admit, having heard only Sondra's side of the story, I was incensed and ready to do battle with that scoundrel Roger. Here, I thought, was a classical confrontation of Good and Evil in the invention business. I would ride in on my white charger, scrag the infidel, rescue the fair maiden, and maybe even get a searing exposé for my next book.

Then I talked to Roger.

It became apparent very shortly that what we had here was not a dynamite exposé, but rather a regular old garden-variety case of crossed communications. I suggested that the two of them get together and talk it over, offering my services and offices as moderation ground zero. (By this time I was fascinated by the whole affair and could no more drop it than you could turn off a good mystery on TV before you find out whodunit.) Roger wasn't all that excited, frankly. Sure it was a good idea, but life is short. Who needs the hassle? Sondra wasn't eager to sit down with Roger, either. With a little persuasion I finally convinced all hands that this marriage might be worth saving.

I tried to explain to Sondra that a royalty deal was the best possible way for her to make money with her ornaments. I tried to soothe Roger a little bit. By the time the meeting was over, an agreement was reached that should have been ironed out in the first place without all the dramatics. It was a good deal all around.

Sondra would do what she does best: have ideas. She would get name credit. Roger would do what he does best: sell products. They would find someplace that could make a quantity of Sondra's ornaments in time for Christmas sales. They had studied the public reaction to the ornaments to find out if the excitement at the trade show would carry over to the cash register in the retail stores. They had eliminated the slow-moving ornaments in the line. And they had studied reorders to get an idea of what sales potential really existed. They were, once again, happily and successfully in business.

THE FOUR ETERNAL TRUTHS

Having now read about some of the fallacies of negotiating a deal with a manufacturer, I imagine you're wondering what the eternal truths are. The first item on the agenda is to take a look at what the manufacturer wants from you in the negotiations.

Fundamentally, any manufacturer you might deal with is looking for the same four things.

First, he wants a good idea. That goes without saying, but that's the first thing you have to convince him of.

The second thing he wants is a good track record. He wants to see that your product is selling, even if it's just in a limited market and only on a test basis. He isn't so interested in the simple fact that some stores have bought the product. That really tells him nothing, other than you managed to make some sales to buyers for the stores. What he's really looking for is a good record of reorders. Earlier in the book, I mentioned that an average product will turn about four times a year at retail. If you can match that, or better it, in a reasonably projectable market sample, you've got a strong selling point.

Okay, suppose you haven't been selling at retail. You haven't got a real-life track record to show him. Then you'll have to make a convincing case with sales projections that make sense, and/or with research results that are believable. One way or another, you've got to convince him that your product will sell. Your enthusiasm is hardly ever enough to convince the manufacturer of that.

Third, the manufacturer would like protection. Either a patent application or a copyright or one of the other things we talked about earlier in the chapter called "Patent It, Fast!" Like Stephen Paul Gnass's choice of the PPA. If the manufacturer thinks he'll be knocked off in the market when he comes out with your idea, he's going to be a lot less enthusiastic about making a deal with you. It's up to you to convince him otherwise.

Finally, your cost figures. He needs to know exactly what he'll have to spend per unit to make your product before he can figure out how much maneuvering room he has to play with in the profit margin. Naturally, he'll do his own cost figures, too. But he'd like to see yours.

YOUR BILL of RIGHTS

What about *you*, the inventor? You've got some things you want from the manufacturer, too. The first, and most important, is a fair royalty. (Or, if you're selling the company, a fair price.) By now, you have a pretty good notion of what a fair royalty ought to be on your product. And if you're selling him a company, you have a right to a fair cashout for your assets—including inventory, equipment, accounts receivable, and maybe even that nebulous but valuable commodity called "good will." And don't forget your intellectual properties such as patents, trademarks, and copyrights.

What about an advance? You've been beating your brains out developing your product, presumably, and it would be nice to have a chunk of money to play with. It would be nice to get an advance against royalties. Needless to say, you may not be able to get it. It depends in part on industry practice, in part on policy at the manufacturer's company. It may even depend on his cash situation. However, as a good friend of mine once said, "You really don't have a deal until some money changes hands." He's right! And that brings us to another thing you want from the manufacturer.

Has he the capital to exploit the market? That is, after all, your best defense against being knocked off. That also implies that the manufacturer has a sales and distribution organization capable of hitting the market hard and fast.

One last thing to think about—primarily if you're selling a company but even if you're just selling your idea: Is there a management contract? If the manufacturer likes your idea, he ought to be interested in any other ideas you come up with. If you're signed on as a consultant to him, giving him the right of first refusal, or maybe even an exclusive right to your ideas, you both do well.

Also, more often than not, you'll do better in your negotiations if you decide to hire an agent. Be prepared to pay between 15 and 50 percent of the net selling price to him, even though he only speaks 150 words on your behalf. I know for a fact most of us cannot be firm enough to bargain on our own behalves. Modesty. Ignorance. Fear. Better you should have a surrogate. Especially if that person likes you and your ideas. By the time you get to the negotiation stage, you'll need someone else. Your opinion of your idea and its value will have already gone down the sewer. You'll be bored, embarrassed, and really not at all ready to be scorned again.

At this point agents earn their money. However, seller beware. Do not, I repeat, do not pay a sales agent an up-front fee. This is another important point Gnass wanted me to emphasize.

And there's only one thing for you to avoid like the plague.

Greed. It'll kill you, if you don't watch out.

WHO'S SCREWING WHOM?

Before I leave this chapter, I'm going to share with you one of the most valuable formulas you'll ever find to cover the agonizing process of establishing a fair advance against royalty. This information did not appear in the original version of my book because I had not learned this very important lesson yet. Incidentally, in the 709 individual licensing agreements I've written, negotiated, and signed, the procedure has always been the same:

"What percentage of royalty do you want?" the licensee asks.

"Five percent of wholesale," I answer.

"Okay," he says. "And how much of an advance do you want?"

I answer by asking, "If you were to have an average sales year for my product, what would be the total wholesale income it would generate to you in the first twelve months of sales following its introduction?"

The licensee then pretends to think for a moment and then he says, "Hell, I don't have any idea."

(Author's note: He's fibbing but you can't point that out and still keep the negotiations on track.)

You then say, "Well, do you think you'll do a half a million dollars the first year?" At this point you get one of two replies:

* "Hell, if we don't do a million, we would not even consider your idea."

* "There's no way we can do more than $200,000 the first year."

Either way the answers go, you're now perfectly positioned to do the licensee's thinking for him.

If the response was the $1,000,000 number, you nod knowingly and offer the following summary: "Let's see, 5 percent of one million is $50,000 in royalties I might reasonably expect were you to have a fair first sales year." Since these are his numbers, he will be forced to agree with you. And at this point you start talking like you just got off the turnip truck. You say, "I think that our cutting that figure in half would be a good start in determining the advance. Let's say $25,000." While he's trying to figure out if you're screwing him or not, you add, "But, since I'm going to make the big bucks off the royalty, let's make the advance half of the $25,000. Let's settle on a $12,500 advance against royalty." In all my existing deals, this is the point where an agreement was usually made.

Could I get more? Probably half the time. Do I know which half? No. And to make this stickie-wicket work for you, you must really believe in your product and the fact that the real bucks will, in fact, come from your earned royalties. And by the way, the same formula works if the number is $200,000. The only problem is you get a lot less money to start.

13

Rubbermaid's
$1,303,620.98

*Here's a real-life example that
proves an important point:
There's hope for us all.*

REMEMBER WHEN EARLY ON IN THIS EPIC I PROMISED I'D GIVE YOU REAL
numbers as regards my successes and failures? Here are the actual royalties
by month that are still being generated by one of my really big successes.

Date	Royalty Paid	Date	Royalty Paid	Date	Royalty Paid
05/86	$ 3,865.00	09/87	18,817.00	06/89	11,595.00
06/86	9,770.00	10/87	19,728.00	07/89	14,267.00
08/86	5,170.00	12/87	21,090.00	08/89	16,204.00
09/86	11,962.00	01/88	19,973.00	09/89	14,310.00
09/86	9,172.00	04/88	39,679.00	10/89	15,448.00
10/86	9,608.00	05/88	10,646.00	11/89	13,239.00
12/86	9,396.00	06/88	20,556.26	01/90	9,361.00
12/86	5,522.00	08/88	21,739.00	02/90	7,592.00
01/87	5,544.00	08/88	21,084.00	04/90	80.00
03/87	9,598.70	10/88	20,252.72	04/90	9,675.00
03/87	6,897.30	10/88	18,782.00	05/90	14,248.00
04/87	10,920.00	11/88	18,837.00	07/90	9,929.00
05/87	9,480.00	12/88	12,156.00	08/90	10,727.00
06/87	12,204.00	02/89	12,962.00	08/90	10,959.00
07/87	15,885.00	04/89	0.00	10/90	10,712.00
08/87	15,430.00	06/89	52.00	11/90	10,621.00

Date	Royalty Paid	Date	Royalty Paid	Date	Royalty Paid
12/90	8,804.00	05/94	6,577.00	10/97	3,225.00
12/90	11,251.00	05/94	5,222.00	11/97	3,121.00
01/91	7,576.00	06/94	6,307.00	01/98	2,103.00
03/91	8,084.00	07/94	8,676.00	01/98	1,969.00
04/91	13,327.00	08/94	5,903.00	02/98	1,929.00
05/91	5,553.00	09/94	9,380.00	03/98	2,245.00
05/91	6,866.00	10/94	8,300.00	04/98	2,738.00
05/91	7,774.00	11/94	4,876.00	05/98	1,978.00
07/91	10,578.00	01/95	5,975.00	05/98	2,612.00
08/91	7,615.00	01/95	4,640.00	06/98	1,631.00
09/91	20,218.00	02/95	5,047.00	07/98	2,797.00
09/91	18,992.00	03/95	4,596.00	08/98	1,692.00
11/91	25,013.00	04/95	6,865.00	09/98	2,318.00
11/91	18,671.00	05/95	5,145.00	10/98	1,927.00
12/91	7,479.00	06/95	5,629.00	11/98	1,518.00
01/92	11,010.00	07/95	6,388.00	12/98	861.00
03/92	9,701.00	08/95	6,329.00	01/99	1,925.00
03/92	10,155.00	10/95	7,649.00	02/99	1,004.00
05/92	15,937.00	10/95	9,451.00	03/99	1,084.00
06/92	11,331.00	11/95	3,019.00	04/99	2,226.00
06/92	10,930.00	12/95	4,106.00	05/99	1,253.00
08/92	13,664.00	01/96	5,450.00	06/99	1,342.00
08/92	12,363.00	02/96	2,071.00	07/99	1,290.00
09/92	10,162.00	03/96	2,650.00	08/99	1,152.00
10/92	12,633.00	04/96	4,244.00	09/99	1,903.00
11/92	9,863.00	05/96	5,177.00	10/99	971.00
12/92	9,008.00	06/96	4,764.00	11/99	1488.00
01/93	7,312.00	07/96	6,858.00	12/99	766.00
03/93	9,196.00	08/96	8,856.00	01/00	768.00
03/93	8,774.00	09/96	6,535.00	02/00	781.00
04/93	5,314.00	10/96	4,460.00	03/00	1083.00
05/93	6,384.00	11/96	4,205.00	04/00	903.00
07/93	7,684.00	12/96	2,594.00	05/00	1130.00
08/93	9,014.00	01/97	5,120.00	06/00	1254.00
09/93	8,733.00	02/97	3,560.00	07/00	1158.00
10/93	10,341.00	03/97	3,596.00	08/00	1179.00
11/93	6,269.00	05/97	3,655.00	09/00	967.00
12/93	6,213.00	06/97	3,606.00	10/00	762.00
01/94	6,770.00	07/97	3,607.00	11/00	761.00
03/94	4,559.00	08/97	3,747.00		
04/94	5,773.00	09/97	4,333.00	TOTAL	$1,303,620.98

Back near the beginnings of my Center for Homewares Design, I was able to convince the folks at Rubbermaid that using our homeware designs would help the sale of their Con-Tact brand shelf liner.

Because of the narrow profit margins on the product line, there was no way Rubbermaid would agree to paying me the 5 percent of wholesale royalty I was asking. However, when they alluded to the potential dollar volume my designs might generate in retail sales, I cheerfully agreed to cutting my then-normal 5 percent down to 2.5 percent. Stick that decision into the back of your brain. It turned out to be a good one, which hopefully you'll get to use someday.

Another thing to remember from this chapter is that you should always have your agreements in writing and have each agreement cover you for the retail life of your product. Also, your agreement should be transferable (as was mine when Rubbermaid sold the Con-Tact brand to Decora Manufacturing in 1998).

The real lesson that only these real dates and real dollar amounts can teach you is that you must always believe in yourself and your ideas. And, for goodness sake, keep persevering. I had already been working at the idea business for over nineteen years before I connected with Rubbermaid on this mini-bonanza.

One-Man **Band** MUST **Toot** OWN **Horn**

14

Using advertising, packaging, and publicity to sell your product.

YOU ARE ABOUT TO ENTER THE MAD, MAD WORLD OF MARKETING and Merchandising. How come? Because you have a product to sell, that's how come. Getting your product into the stores of America is only half the battle. Now you're going to have to move it *out* of those stores and into people's homes. That takes a lot of effort.

Marketing means different things to different people. To me, marketing is a circle made up of equal doses of product development, manufacturing, financing, sales, merchandising, and management. In my opinion, no single part is more important than another.

Advertising, as most of us understand it, is part of the merchandising picture. But for a new inventor trying to get his new product going, advertising is the least of his worries.

So many new inventors I've talked to have grandiose plans for advertising their products. They talk blithely about "taking an ad in a magazine" or even "doing a television commercial." Most of them are shocked when they find out the costs involved in doing some of those things. Just for your information, one full-page, four-color ad in *Time Magazine* will cost you around $192,000 just for the space alone. That's not counting the $25,000 or $30,000 it will cost you for production—getting the pictures, doing the mechanical layout, and getting the production work done.

And that's just for one shot at it. Any advertising campaign worth its salt comes back several times in a number of different media. For General Motors or the telephone company, that's just part of the cost of business. Big-time advertising is *not* the topic of this chapter.

However, once you reach the big time, you'll discover that most companies allow about 3 percent of their total sales for advertising. To introduce a new product is worth an absolute minimum of 10 percent of sales. In the case of General Motors, that gives a pretty substantial budget for advertising media and production and creative service.

At the beginning of a product's life in the marketplace, there are better ways to get results than to spend your time working on retail consumer advertising. We're going to address ourselves to some of the more practical aspects of merchandising. The subject matter of this chapter, then, includes how to:

* Design a package so it sells the product.

* Make your product sell itself to the public.

* Get someone else to pay for your advertising.

* Use "collateral." And what is it, anyway?

* Knock yourself off for fun and profit.

* Put Hollywood to work for you.

YOUR PACKAGE is your FRIEND

Let's start with the most basic advertising medium you have in your arsenal: your package. If you think of the package as just something to hold the product, you're missing out on what may be your only real chance to give the public some sell copy. Remember, you're going very light—if at all—in consumer advertising. At best, you've got a little trade advertising, some word of mouth, a few odds and ends. Most of the public will learn about your product for the first time when they see it in the store. At that point, the most important function of your package is not to hold the product, but to serve as a billboard for advertising your product.

That being the case, your package had better be capable of fulfilling that function. There had better be enough room there to do a little selling. For example, if you are using a "blister pack"—a clear plastic bubble or "blister" on a card holding the product—make sure the card is big enough to hold some words and pictures in addition to the product. If your product is in a box, so much the better. A box is a five-sided billboard, with each side selling your product. Make the most of it. The side you don't use to sell with is the one it sits upon!

Think about how your product will be displayed in the store. If you're using a blister pack, it will be hanging on a rack of J hooks, probably, so the front of the pack is the only working part. You'll have to use it for all it's worth. If the product is in a box, it may be stacked on a shelf so that only the end is visible. Or maybe the side. It won't matter that the top of the box is a masterpiece of advertising design if nobody ever sees it until it's been bought. Even some of the most seasoned advertising professionals in package design have been known to forget this important little fact. Use all the sides of the box so that each one can stand on its own as a billboard for your product.

The thing to remember is that your package has an awful lot of work to do in a terribly short time. If you're very lucky, the customer may stop to look at your package for as much as four seconds. That's all you can reasonably hope for. Four seconds is all the time you've got to convince him that he needs and wants what you have to sell him. In that short moment you've got to tell him what your product is, what it's called, what it does, and why that's a good thing.

With a big four seconds to play with, you're obviously going to have to keep it simple, keep it uncluttered, and keep it direct. Show him a picture of the product in use or in some setting that gives him some information about how it works and how it's used. Keep the words down to a minimum, but make sure they're not so terse that they don't make sense. He's not going to stand around reading *War and Peace* on your label but he *does* want to learn something about your product. Otherwise he wouldn't have stopped to read what you have to say. Tell the person quickly and concisely.

In the beginning of the Rickie Tickie Stickie saga, I chose the cutesy-pooh name Rickie Tickie Stickies and then set it in a totally illegible typeface of my own design. That was wrong, wrong, *wrong!* By doing what I did, I flew in the face of conventional wisdom. I *should* have been wrong. It was just a fluke that I turned out to be right. The rule is that you're much better off with a simple name on your package, especially if that name gives the buyer some inkling of what your product is and what it does. And when you put that simple name on the package, for God's sake, make it readable. That's what conventional wisdom says. And usually it's right. I guess what that means is that what I've told you so far is correct, but maybe not necessarily right for your case. All I can give you is the best thinking I've been able to come up with. If luck and a fluke of circumstances turns all that wisdom inside out—who am I to complain? I benefited from that kind of fluke, too.

Any advertising piece—including a package—has to tread a fine line between being lovely and being informative. As the client, in the case of your product, you'll have to see to that balance when you're dealing with the advertising experts you hire to take care of your design needs. Your package ought to be attractive, certainly. But it doesn't have to be a museum piece, if that gets in the way of its function. On the other hand, it ought to work hard at selling, but it should not look like an ad for a fire sale. It's a delicate balance. All I can tell you is that in a standoff between the two claims, "informative" gets the nod over "pretty" every time. Usually, though, it's possible to be both.

POP GOES the WEASEL

One of the best, most effective ways to increase the sales of a product is to provide the store with a display unit that shows the product to advantage while it does some selling of its own. It's a way to help your product help sell itself. This may be a floor-display unit, or it may be a display designed to stand on a counter. It might be a multisided rack or any number of different kinds of display units. Early on in our Stickies adventure, we discovered that by using a floor display unit we had designed, we could count on selling forty or fifty percent more Stickies. (That was a special case, and you can't necessarily project those figures to any display unit. But you can be sure that displays do improve sales to some degree or other.)

Paperback book publishers have been making use of this bit of knowledge in recent years. Many of them now offer "dump bins" (a form of display unit) to bookstores as a means of displaying their current crop of new titles. These displays take the books out of the anonymous ranks of book spines on the store shelves and put them out in the aisles or on counter tops where they can be seen and where the cover design can help sell the books. It's one of the few instances of real merchandising moxie in the publishing business, which seems to be living in the Stone Age of merchandising in most other respects.

Usually a display unit is shipped at no charge to a retailer as a reward for making some larger-than-ordinary order of a product. And it is usually designed in such a way that the order is shipped already packed in the display. That saves shipping cost, which, in the case of a standing floor display, can be considerable. There are people who make a nice living designing and selling these displays, and you'll do well to get in touch with some of them. If your order is going to be reasonably large, they may design your display unit as a part of the box order. Otherwise, you may have to pay a designer's fee. Either way, this is not a job for an amateur.

Almost without exception, these display units have what's called a "header card" or an additional piece that mounts on top of the display and serves as a billboard for the product. The construction of the display can range all the way from plain cardboard to heavy corrugated cardboard to (somewhat rarely) wood and glass and aluminum. It depends on how long you expect it to last and how much you've got to spend. If it's a permanent unit, you may spend quite a bit of money on it—some of which you may recover from the retailer, since it helps him sell, too. If it's strictly for a short-time promotion, it won't have to be built like the Grand Coulee Dam and won't cost as much, either.

A full color, floor-standing corrugated display unit will usually run between $15 and $25 each in quantities of 1,000. To that, add another $5,000 for design and production art and you're looking at a serious investment.

The guidelines for designing a display unit are about the same as the guidelines for designing a package. Make it attractive, informative, and efficient in getting the message across. Find the single most important thing you can think of to say about your product and say it in a way that gives it impact. There's still only about six seconds to get your message across. The difference is that with a display setting your product apart from the rest of the stuff in the store, you are more likely to be noticed.

I must tell you, by the way, that there are an awful lot of places where you can't use a display because of store policy. Most of the better department stores won't allow floor displays or counter displays of any kind other than the ones they themselves have put in the stores.

Their theory is that they have spent a lot of money and effort to create a "buying atmosphere" and a "personality" in their stores. Anything that isn't a part of that carefully calculated decor doesn't belong there, even if it increases sales for a specific item. For that very reason, by the way, some department stores and chains around the country have begun branching out into discount operations, usually under a different name. This permits them to participate in the rack-and-display, self-service atmosphere of high-volume sales without sullying their uptown image.

What it all means to you is that you ought to examine very carefully the kinds of stores you plan to be selling in before you invest in display units. If hardly any of them use displays, you'd be wasting money. Also, even if there's no set store policy against displays, you may find that many stores don't use them anyway. Even the biggies get caught in that trap from time to time. Bristol-Myers, for heavens sakes, did it once. They spent a lot of money with me, an absolutely *inordinate* number of dollars, on ten thousand standing display units. They were beauties, too. And they sold the product effectively,

151

besides just looking handsome. Bristol-Myers gleefully shipped 'em out to their retailers and considered it money well spent—until somebody did a survey of the stores. About one of ten displays was getting used. The rest of the investment was gathering dust in storerooms or had been given the old heave-ho practically on arrival. There was severe chagrin at Bristol-Myers.

Your best insurance is to call some buyers at the stores you plan to be selling in, or at least the type of store you plan to be selling in. Ask them for the real lowdown on whether or not display units get used. They'll probably be happy to tell you, since it saves them trouble, too.

CO-OP ADVERTISING

Let's move on to a somewhat more traditional form of advertising—with a twist. It's called "co-op" advertising and it looks, feels, and smells just like regular media advertising for your product. Newspaper ads, maybe radio. Possibly TV, even. The twist is that someone else pays for most of it. You ought to be aware of it because it's a beautiful way to get a lot of mileage out of a practically nonexistent advertising budget. There are a couple of ways you can work it.

We'll take the simplest way first. If a customer is ordering a lot of your product—bigger numbers and more regularly than the average run of your customers—you cut him in on your co-op advertising allowance. What that means is this: Every time he orders from you, he gets to deduct a set percentage of his purchase from your invoice. He uses that deducted money to help pay for advertising in which he mentions your product. Commonly, the advertising allowance is 10 percent of his invoice total on any given order. So if he orders $1,000 worth of product from you, he actually pays you $900 and spends $100 on advertising for your product.

What's the catch? None, really. Except that the retailer who's buying from you won't be very interested in advertising your product if it isn't likely to bring people into his store. Therefore, you have to have established some momentum for your product before co-op advertising makes any sense to the retailer. By the same token, you should be getting tear sheets or other proof of the advertising he is running, just to be sure it really *does* mention your product. (In the case of radio or television, he should send you a tape and a schedule of where and how often the commercials are appearing.)

Obviously, you have to set some sort of standards for what constitutes a mention of your product. Two words do not a mention make. But if you're reasonable about your demands, the retailer will usually give you a fair shake in his advertising in return for your co-op allowance. There's really no way to set up hard, fast rules on a co-op advertising program. It's all pretty much a matter of judgment and trust between you and the customer. Sometimes that can be a screeching pain. Mostly, though, it's a pretty easy system to deal with, and it does both of you some good—it helps out his advertising costs appreciably while it gives you a lot more exposure than you could get on your own.

The next level of co-op advertising is a bit more complicated. But not much. It's the same deal as before, but in this case you also supply the retailer with some of the advertising material he'll use. You make up some newspaper ads, maybe some radio commercials. Perhaps you'll even do some television, if you're feeling particularly affluent and if it makes sense in the light of your customer's advertising plans and your product.

In each of the ads, you feature your product heavily, but you leave room for the local merchant to put his name, his store location, and so forth. And you build the ad in such a way that it's your customer who's talking to the public, not you. It's his ad, not yours, even though it's all about your product.

You provide the ads already printed on a slick-surface paper or on a disk. These preprinted ads are described as "camera-ready" and they are called "slicks." The disks are called "disks."

In the case of radio commercials, you can do any of several things. You can provide fully produced radio commercials—announcer, sound effects, music and all—on tape. These commercials will have a "hole" left in them for the station announcer to fill in the local retailer's message. Or, you can provide preferred "tags" of five or ten seconds. The retailer drops these tags in at the end of the commercials he is putting on the air. You might just send scripts for five-or-ten-second tags to be read "live" by the local station announcer at the end of the retailer's commercials.

The main thing to remember is to give the retailer as much leeway as you can in how he uses your material. In print advertising, give him a selection of different sizes, based on standard column widths. You might give him a one-column by three-inch ad, a two-column by six-inch ad, and a three-column by ten-inch ad, for example. And give him little "plug-in" modules to use as a part of a multiproduct ad, too. In the case of broadcast material, give him different lengths and different formats so that he can fit you into what he's planning to use. This whole area is one where an amateur could really get himself into trouble and waste a lot of money. Hire, or better yet, involve an expert and let him do his thing. Just be sure the message is clear and has impact. Leave the technicalities to experienced people.

By the way, there may be special cases where you throw the guidelines overboard and give a much bigger co-op allowance on a one-shot deal than you normally would. I ran into a good example of that with the Stickies. A big store in Colorado—at the height of the Rickie Tickie Stickies madness— ran a full-color, full-page ad in a Sunday supplement magazine and the ad featured our Stickies; the store was building a promotion around my flowers. The page cost the store, in those days, $1,500, and I agreed to split the cost right down the middle, even though they had only ordered $2,500 worth of my product. I figured it was well worth the exposure to spend that kind of money in this case. And I was right. The store sold out almost instantly, and the reorders kept rolling in, thanks to the momentum given us by that store promotion and the ad that kicked it off.

GOOD COLLATERAL PAYS

"Collateral" is a word you're going to run into sooner or later when you start merchandising your product. Collateral is advertising and merchandising material that doesn't go into regular advertising media like newspapers, radio, television, or on your Web site. It includes things like catalog sheets, store-display materials, and even things like business forms.

Putting together a good, solid collateral program is pretty basic to your whole merchandising effort, even though much will be seen only by the trade—the salesman, the retailers, and the other people whom you have to reach in order to get your product into stores. That's only half the battle, as I said before, but it's the *first* half of the battle.

Let's take a look at some of the more important parts of your collateral material. First and foremost is your catalog sheet. This is a single-page, sometimes full-color sheet of paper that goes into a three-ring binder used by your rep when he's taking orders for the products he sells. It explains what your product is and shows whatever different models you have to offer. Doing the catalog sheet is kind of a litmus test of an idea. I've found that if I can't explain a product on one sheet of 8.5″ × 11″ paper, it's probably not a good idea. Naturally, you'll want to show your product in its best light, with handsome photos and/or illustrations and with words that not only explain it but help sell it.

Do *not* print your prices on the catalog sheet. Prices go on a separate, one-color sheet. That's because they may change during the year. The way things are going these days, the prices will undoubtedly change. Upward. For a number of reasons, you want to be able to change the price list quickly and easily without having to redo the whole, expensive catalog sheet. You ought to figure on going through 5,000 or more catalog sheets per year. I don't know where they go, but they somehow get used up in large quantities over the course of a year, so you might as well go ahead and order a bunch of them at once to take the printing discount that goes with larger orders. All of your reps will want several of them, of course. Some of your retailers will want to keep then on hand. You may mail them out to retailers yourself to stir up some interest in your product. And you may give them out at trade shows, too. You'll need all of them you print, believe me.

When you're printing up your business forms—invoices, order sheets, stationery, and business cards, for instance—you might as well put it all to work for you. Plan to design it all so that your product is featured on it in some tasteful form. Maybe you've got an advertising line or a slogan of some sort that has been effective or that you like a lot. Put it wherever it seems to be appropriate. No, you don't want to turn your correspondence into an advertising extravaganza, and you certainly don't want a tacky letterhead, but it can't hurt to plug the product somehow.

There is also a gray area of collateral about which you'll have to make your own decision, based on your product and your market. You can do things like window banners that plug your product, if you think the stores where you'll be selling might use them. They can be just about any size, including jumbo three or four foot babies. If you do window banners, make sure they can be read from a car passing in the street. The pedestrian is becoming an endangered species in America, so most of your passersby do it in a car. And for heaven's sake, check with your retailers before you plunge ahead and run off a batch of banners. They may have size restrictions you don't know about, and they may have other quirks that could make your banner worthless to them.

Also in this gray area is something called a "shelf talker." That's a sort of tag or display piece that attaches to the shelf where your product is displayed. Generally they're small—no more than three to six inches wide and maybe one to two inches high, depending on how and where they'll be used. The shelf talker catches the attention of someone walking down the aisle in a store and might make him stop long enough to give you that six-second shot at him with your package. They do work. You just have to be sure that the stores where you sell can and will use them.

There are any number of different kinds of items like these shelf talkers that you can use. They all fall under the general heading of "point-of-purchase" material. (The name is self-explanatory. And it's a fact that merchandising done right at the spot where the person makes the choice of buying or not buying seems to work very efficiently.)

Gary Dahl, the father of the Pet Rock, had a stroke of genius and turned a piece of point-of-purchase material into a lot of sales for his baby. Since this little beauty was probably way before your time, suffice it to say that it was one of the most brilliant new-product ideas of the 1970s. Or 1980s. Or ever. A rock in a pet carrying-case box with an obedience instruction book? For $4.00 retail. Beautiful. The Pet Rock hit its stride in the pre-Christmas period so that lots of dealers stocked up extra heavily for the holiday. Naturally, there were some left over at the end of the Christmas season. Gary was very interested in keeping the ball—or the rock, in this case—rolling, so he designed hang tags to go on the clever Pet Rock carrying case that serves as the package for his product. The hang tags neatly positioned the Pet Rock as the perfect Valentine's Day gift for loved ones with a sense of humor and a need for an easy-care pet, like Gary's purebred igneous rock. Suddenly all the retailers had a whole new selling season for the Pet Rock.

Naturally, Gary was able to do the very same thing for every holiday from St. Swithin's Day to Tisha B'Av. Nothing about the product or the package changes. All it takes is the addition of an inexpensive piece of point-of-purchase advertising to the original product and your Pet Rock is in position for any special event on the calendar.

KNOCK YOURSELF OFF

At various points in this book, I've been referring disparagingly to knock-off artists. Now I'm about to do a full about-face and discuss how you can become a knock-off artist. The target: your own product. If it sounds wacko, stick around. There's some very serious method to this particular stroke of my madness.

Remember back when we were discussing how to price your product? We talked briefly about how a price that was too high—or too low—could be a problem to you. Well, there's no changing that, but you *can* change something else in that equation. You can change your price. You can knock yourself off by marketing a cheaper version of your own product so that you can fit neatly into other markets.

The first rule of this game is that you have to enter the market at the high end. Make your deluxe model with the fancy packaging and get that rolling. Once that's underway, you build a somewhat cheaper version of the product, if possible. If you can't do that, at least you can do a cheaper package and take a slightly smaller profit per unit to get yourself into a lower price point but higher-volume market. You can make as many steps down as you can find elbowroom for in your price. The only rule is that you have to start at the top and work your way down. It's very difficult, if not impossible, to start with a cheap version and then market more expensive versions in the classier stores.

While you're doing this, you have to market your product under different brand names, of course. Robinsons-May wouldn't be very happy to find that same brand name and the same product was selling at Rite Aid for considerably less, in a more plebian package. But if you call the Robinsons-May version "Acme" and the Rite Aid version "Amalgamated," you aren't likely to stir up animosity at either store. Retailers expect manufacturers to get knocked off from time to time, so they won't be surprised to see that competition has developed at a lower price level. Since Rite Aid doesn't compete with Robinsons-May, no serious harm is done. They both simply address their own individual markets with their version of the product, and everybody's happy. There's nothing illegal about it. I feel it's just plain smart. Your goal is to make a success for yourself, so why not cover all the bases?

The advantages of knocking yourself off are pretty clear. The most basic is that if you don't, someone else probably will! Selling your product at a classy store pretty much cuts you off from a whole market at other price levels. Staying upscale will cost you about 90 percent of the market potential for your product. By knocking yourself off, you have access to the lower level, too. Your sales penetration increases, and your profits go up. So if there's enough of a profit spread to work with, you can give sales an enormous boost this way. You can hit intermediate steps in the price scale, too. Just be aware that by going from a swanky department store to a discount chain you are getting into a whole different distribution structure, which may have an effect on your pricing. Read over the chapter on "Which Way to Go" before you get too far into knocking yourself off or you may find that you have knocked yourself *out*.

TOOT. TOOT. TOOT.

One of the most important areas of merchandising is one that all too many new inventors ignore entirely. Publicity is usually better than the best advertising. For one thing, it's practically free. For another thing, it usually gets the kind of audience you couldn't buy for any price. And finally, it is more effective because people are more likely to read and believe something in regular news media than they are in an ad. Take it from me, publicity is worth its weight in solid-gold ads.

Here's where that one-person band starts tooting his or her own horn.

The very first step to take in generating publicity is to spend a few hundred bucks for a first-class, professional photographer. You will probably be needing photos of your product for your package, your catalog sheets, and other purposes anyway, so you'll need to spend the money no matter what happens. With a set of good photographs in hand, you have the first and most important element of a publicity campaign. Whenever possible, include yourself holding your product. Very close to your chin. Seriously.

The object of your exercise is to get a story about your product published, so the next thing to do is figure out where you'd like it published. An obvious place to think of is the trade press. There are all kinds of magazines published, each catering to some very specialized audience. Everything from the *National Nurse Anesthetist Journal* (try saying that three times, fast) to *National Knitted Outerwear Times*. There must be at least one trade publication designed to serve the kind of retailers who will be selling your product. Find out the name of the publications that reach your trade audience. How? As I keep telling you all the time, ask. Call a store you think will carry your product. Ask for the buyer of your kind of item (as Mrs. McDonough did with her trash separator). Ask that buyer what trade publications he reads. Then go after them.

Don't stop there, of course. Another natural for you is a newspaper wire service, like the Associated Press. These services are always on the lookout for a good feature story—something with a newsy slant but basically just an entertaining or informative article. Think also of other publications you'd like to be in with your product. In our case, with Rickie Tickie Stickies, a natural place to try was the *Los Angeles Times* Sunday supplement, called *Home* magazine. Our flowers were sort of free-spirited and funky, which matched the way *Home* magazine viewed the Southern California scene. It made sense. In the case of your invention, you can pick out equally appropriate publications.

The next step is to figure out a story angle. Why would the publication run a piece about your product? What's in it to interest their readers? Those are the very same questions the editor has to ask himself, so put yourself in his place. Figure out what kind of story he'd really like to have and his readers would really be interested in, then see what you can do to fit your product neatly into that kind of story.

If it looks as if your product were dragged into the story, kicking and screaming, just to get some publicity, you won't get into print. Stay away from far-out angles. Look for a good, solid, realistic, interesting slant and build the story so that your product is a logical part of it.

Once you've figured out a good angle for the story, do a very brief outline of the way you see the story. Type that up neatly and send it, along with a good selection of photographs to match the story, to the editor of the publication you have in mind. Include your address and phone number so that the editor can get in touch with you for more information, photos, or whatever else he'd like for the story. And don't expect to see any of the pictures again, whether the story is used or not. Editors are swamped everyday with press releases. They can't afford to spend the time to return them. It's just part of your cost of doing business, like so many other things I've discussed in this book.

If the publication uses color photos, send them nothing smaller than 4″ × 5″ transparencies and never send color prints. Small transparencies and prints on paper give the publication problems in production, so they give the editor an excellent reason to forget about your story. If you send black-and-white photos, send top quality 8″ × 10″ glossy prints. The kind you get at the drug-store or the overnight photo places tend to be fuzzy and not terribly well developed. Spend a little extra to have your photographer make studio-qual-ity glossy prints.

I said earlier that you'd be needing photos for a lot of purposes, so you should get good ones that can be used for publicity. But if you are aiming at a big, important, general-circulation magazine like a Sunday supplement, you'll probably want to set up a special shooting session, either in a studio or on location somewhere, to make special pictures that match the story slant. That sounds like a big deal, but it's not as expensive as you might think. About $800, maybe $1,000, judicially used, should get it done for you. Especially if you can get models who are not professional (but who look pro-fessional). When we were shooting some of the Rickie Tickie Stickies stuff for our package design, we used neighbors and friends and their kids. We got signed releases from all of them, in return for a dollar, and we had a supply of soft drinks and snacks on hand at the session. Later, we sent all of them nice sweaters as a follow-up gift. They had a good time and enjoyed seeing themselves on the package in stores, and it all cost a great deal less than using professional models. (The signed releases are very important, by the way, even if you use professional models.) The releases can keep you out of trou-ble later on. Especially if your product takes off like wildfire.

The benefits of a little effort in the area of publicity can be astounding. We got our story placed in *Home* magazine partly because we had done a pretty fair job of getting a good story together. But it was also partly because of a fluke. What I didn't know at the time was that the editor was in desper-ate straits. A story he had been counting on for the next issue of the maga-zine had fizzled out on him, and he was looking frantically for another good story to replace it. It had to be good, of course, but it had to turn up pretty soon or he would have a lot of leftover space in the magazine.

Our little story came to him like a gift from the gods, and he snapped it up. We got a two-page, full-color spread featuring our little flower as part of the new design scene in good ol' easy-living Southern California. It was incredible publicity for us, of course, and it was a good, colorful, appealing story for the magazine, too.

Then the ripples began to spread.

Editors and associate editors of newspapers and magazines read all the other magazines and newspapers they can get their hands on. It's part of their job, scouting story ideas, keeping up with what the competition is doing. Somebody at the *Chicago Tribune* saw the story in *Home* magazine and thought it was a gas. So, several months later, the *Chicago Tribune* Sunday supplement had a very similar story on our flowers. The paper called us, we sent the *Tribune* the pictures they wanted, and we had our first spin-off story.

The next thing that happened was that one of the buyers at Marshall Field, the big Chicago department store, saw the story and liked the flowers. He placed a big order with us and we gained a solid foothold in the Chicago market.

Meanwhile, we had managed to stir up interest at UPI, and the next thing we knew we had a nice feature story, with a picture, on the UPI national newswire. By the time we quit counting, we had seen that story run in more than two hundred newspapers all over the country. What happened next, I'm still not sure I believe.

Whether it was the UPI story, the *Chicago Tribune* Sunday supplement feature or the *Home* magazine piece, I don't know, but somebody at CBS in New York got interested in Rickie Tickie Stickies. One thing led to another and the next thing we knew we had a network news crew dragging cables and lights into our living room to film a segment for the Walter Cronkite news. We got something like seven and a half minutes of national network coverage out of it—something no amount of money could ever buy. You couldn't even match that kind of exposure with an hour of network commercial time.

Years later we had the same kind of experience, parlaying a feature piece into a whole series of media events. This time the product was those Parent Protest Posters I mentioned in an earlier chapter. Nearly thirty months after we hit the market with the product, an editor at *Newsday* on Long Island in New York saw the product in a store and got in touch. To the editor, it was a brand-new product that she had just discovered. Did we have any pictures and information she could use to do a story? You betcha, ma'am! So we got a nice, long story in *Newsday*—a very highly regarded newspaper in those days—about our posters. It was great exposure in a very important market area. Better still, the *Wall Street Journal* was working on an article for its widely read front-page daily feature. The subject? New products on the market. The *Journal* picked up details from the *Newsday* article, and we found ourselves on the front page of the *Wall Street Journal* all over the country.

Then the producers of the *Today Show* called us. They had seen that piece in the *Journal* and were fascinated. Would we mind if they did a number on us? No! And there we were again, on national network television.

And that led to my doing *Oprah,* which in turn lead to *Nightline.* That got me on *Regis* (when he was still in L.A.). And so it went.

As you can see, if you do a good job of generating publicity, you can get benefits you never dreamed of. Maybe you won't wind up rubbing elbows with Walter Cronkite and Barbara Walters, but you could sure get a lot of coverage by important news media.

One of the smart things we did was to hire a news-clipping service as soon as it became apparent that we were going to be showing up in a lot of different publications. For a relatively small fee, the service clipped out every story that mentioned us, and they sent all the clippings to us. Later we were able to use those clippings to demonstrate the kind of interest we had stirred up across the nation. It was a valuable sales tool that helped us cash in on all that publicity.

HI. I'M GEORGE W.

I've just finished doing a little name-dropping. You can, too, with a little bit of effort. If your product seems to be working out well, you may find that you can get an endorsement from a big-name celebrity to use in your merchandising efforts. I mean a *big* name. You may not have the cash to afford the celebrity's usual endorsement fee, but by doing some calling around to agents, you may be able to work a deal anyway. You could pay the star by cutting him in for a percentage of your profits on the product. A lot of big names might find that a very attractive deal for tax reasons. If they can find a way to spread their earnings out over a few years and avoid taking a big lump sum for their services, they're usually interested. After all, fame is all they have to sell, and fame is a perishable commodity. Stars and celebrities tend to make the bulk of their money in a span of a few years, which means that the Internal Revenue Service gets a very healthy portion of it. By deferring or spreading out that income, they can save some of it from the eager clutches of Uncle Sam. If you offer them a way to do that, you might be able to attract one of the bright stars of sports, television, or Hollywood to endorse your product.

DRESSED to KILL

There's another way you can put Hollywood to work for you. The people who make movies and television shows constantly are faced with the problem of "dressing the set." They have to put props into every scene to help the illusion of realism along. (Hardly any "room" you see on film or TV is really a room. Almost all of them are sets, built on a sound stage and filled with props.) Naturally, they can't keep using the same tired, old props, so they're constantly on the lookout for something new to use in the scenes. Your product might be just the thing. If people buy it and use it, then it is presumably a real part of every day life. Anything that helps the film people portray everyday life is all to the good.

You might consider sending a sample of your product—or at least a photo and some descriptive material about it—to the three television networks and to several of the major film studios.

Ideally, you would send your material to the art director in charge of each individual series currently in production and each feature film currently in production. That may not be possible for you, since it involves keeping up with the trade papers to see who's producing what and where. So, less ideally, you would send your material to the vice president in charge of production at the three networks and the studios.

Every art director in film keeps something called a "production book," which is a loose-leaf binder full of sheets listing sources of things he needs in making a film. Find a way to fit your information for him on a standard three-ring sheet (your catalog sheet is a good thing to include) and send it to him in that form. Enclose a note saying that this is for his production book and that you'd be happy to cooperate with him if he ever needs some of your product for a production.

You may strike out at every one, but then again you might find yourself on television or in the movies. I sent samples of Rickie Tickie Stickies all over Hollywood and they turned up on a number of TV shows that I'm aware of and probably some I missed. They weren't necessarily prominently displayed, but they were sure there on the flickering silver screen. Every little bit helps.

And as for the boffo ending this chapter deserves, check this out. I was paid a $150 use fee by the producers of the film *Cocktail* so that they could use a copy of the original version of this book. Tom Cruise was reading it behind the opening titles.

15

Show Me
THE Money

*Now there are two
simple solutions to your
cash-flow problems.*

I WISH THE CURRENT PHENOMENON COVERING HOW MOST SMALL retail stores now pay for the merchandise they buy had been in effect in the bad old days. I probably would have been able to sell $6 million worth of decal products instead of the $1.3 million I was able to haphazardly finance during my first year in business.

And what are those swipeable magic words now available to small manufacturers? Would you believe *credit cards?* Believe it. Today, most small stores now prefer to pay for merchandise with the card they owe the least to.

Think about how that could have helped me then, and you can immediately see how this fact can help you now. All of a sudden I'd have had my 6,800 little accounts paying me immediately whenever they ordered. I would have been my own bank! Don's Bank has a nice ring to it.

Just writing about this amazing change in the current financial seascape is giving me an upset stomach. But that's my problem, not yours. Enjoy the new way of doing business. It is a great boom to the little guy. Especially the little inventor who is determined to take his great new product to market by himself.

If, on the other hand, you are by now a true believer and you only want to license your idea or invention to a manufacturer, do I have a deal for you.

DO I HAVE a DEAL for YOU

I have alluded to the fact that a good way to finance the development of your idea or invention is to make a series of arms-length agreements with the various talent you need to bring your product up to The Velvet Presentation Mode. Now, I'm going to lay it out clearly and directly. Most of the professional inventors I know have all used this approach when they were getting started.

Let's suppose your idea involves your needing some top-quality artwork. The first place you start is the yellow pages under the heading Top Quality Artwork For Sale. Visit the person with the largest ad. Show that person what you need and ask for a written Production Cost Estimate (PCE). As a rule of thumb, you'll be expecting to pay about $100. The PCE will usually come in at about $1,000. So what to do? Talk to the art person and explain what you're doing. (You did get your legal protection up and running, didn't you?) Convince that art person of the fact that all you can afford at the moment is $100.

You would be willing, however, to write up an agreement that pledges that the first $900 you receive from your advance against royalty will go to the art person. And, furthermore, you will continue to pay the art person (out of future earnings on your great new product), small amounts each month until that person has been paid a total of $5,000. With the $100 available to cover his out-of-pocket expenses, the art person will only be risking some time. And your $5,000 possible return on a $1,000 crapshoot is almost irresistible. Especially to the arts-and-crafts types.

Keep in mind that the above formula will work most of the time on almost any of the tradesmen you might need for developing a truly professional presentation. In that group are market researchers, artists, writers, prototype makers, and photographers. Patent attorneys should be in that receptive group, but generally they choose to remain above the fray and bank their fees.

THE IDEA BIZ

And while we're at it, let's figure out how you can cover the patent application costs you will be paying the attorney. First of all, I hope I got stuck in your head the fact that I want you to open a docket and complete your search. That must be bought and paid for every time. Generally, that's where I stop until I get some interest from a potential licensee.

But that still leaves you with having to cover about $1,000 in legal work. And you have only one $100 lump left. What to do? Syndicate. Sell off a 10-percent piece of your idea to someone you know who is willing to invest $5,000 for the pleasure of being in The Idea Biz. Now if $5,000 from one person is impossible to come by, then try getting $1,000 each from five people. Don't increase your group beyond ten, or you will be on your way to becoming a scam artist and the SEC will look well beyond askance.

There's one last point I want to leave you with before we move on. You'll recall that I created and taught an extension course at my alma mater, UCLA. It was called "How To Get Your Idea To Market (Without Giving Up Your Day Job)." The most-often-asked question during my all-day seminars was, "How do I keep my boss from finding out that I'm working on a great, new-product idea in my spare time?" That question always amazed me because it was 180 degrees off the point. You want your boss to know you're a part-time creative genius for two distinct reasons.

First, you'd be amazed how few people actually reduce their ideas to practice. And that fact alone also sets you apart (in your boss's eye) from your other fellow employees. Having an idea (and doing something about it) clearly establishes you as a creative and resourceful person. That you told your boss up front also ascribes to your integrity. It also creates the bond of shared knowledge that can't hurt any boss-employee relationship. All you do when you advise your boss of what you're up to is assure him or her that 99 percent of the time your project will be moving along outside the workplace. Were any work time compromised, you will make it up double and at no salary. That's fair. And I'll bet your boss will accept your proposal and appreciate your involving him or her in your project.

Second, by sharing with your boss, you now are on the line to someone outside your family that you're going to get your great, new idea to market. Added pressure never hurt.

Also, scrupulously keep track of any personal phone calls, e-mails, or faxes you might have to use on company lines. Pay your accounting department each month even if it's only pennies. It is amazing how that little act of honesty will also set you apart from most of the other employees.

GOOD NEWS. BAD NEWS.

For those of you who absolutely won't take my advice about licensing your invention and insist on becoming a boot-strapping entrepreneur, then this next part, which is titled "Who's Got The Money" and is lifted verbatim from my original book, will give you invaluable advice and insight from my own experiences.

There are times when it seems you have to be rich to begin with if you're going to make a success of your invention. There were times at the early stages of our adventure when I was convinced the whole thing was going to die from acute fiscal anemia.

All the money you have to spend! In the last chapter, we talked about getting people to design packages and display units, hiring copywriters and art directors to do your advertising and collateral material, not to mention photographer costs and models. Then there's the cost of all those 4″ × 5″ color transparencies and 8″ × 10″ black-and-white studio prints you'll be sending around in the hope of generating publicity. If you wind up manufacturing the product yourself, even as a market test, you'll face the cost of tooling, manufacturing, packaging, and distribution. Even getting a patent costs money.

Where does it all come from?

Well, there's good news and there's bad news. The good news is that I'm going to tell you how to get a lot of work done without paying any cash. The bad new is that you'll still have to spend some big money. Worse, there's a financial problem I haven't completely explained yet.

HERE'S why I LIKE LICENSING

The new financial problem is something I call "pocketbook lag." The pros call it cash flow. To illustrate, remember how I told you I sank $1,000 into that first batch of Stickies? (That's not counting the photography costs for our publicity pictures and package design, nor does it include the cost of the special dies we had to provide the printer so he could make the Stickies for us.) To get that money, we had dipped into the vacation slush fund that we figured we could afford to risk on this new idea.

Shortly we sold out all that first batch of Stickies and had a new stream of orders coming in. It was time to go back and print up another batch. This time it was going to be a bigger order, too, so we could at least break even on the printing costs.

That's when we discovered our principle of pocketbook lag. The money hadn't come in yet from those first sales. You see, in business a really good account is one that pays in thirty days. It's not uncommon to wait sixty or ninety days or even longer to be paid for your product. In fact, the area of business in which we were operating—gift and stationery and department stores—is notoriously slow-pay. At the same time, suppliers expect to be paid cash up front, especially when they're dealing with a brand-new business like ours. That's pocketbook lag.

We needed $2,500 more now. Hitting the vacation fund was one thing, but now we were going to have to go into hard savings, and that was painful. Still the idea was giving indications that it was going to pay out well, so we bit the bullet and spent the savings.

Just about that time, though, we hired a rep and our publicity efforts began to pay off. On the basis of his contacts in the trade and the fact that we had a story about to break in *Home* magazine, our rep got orders from five hundred more stores around Southern California. The ink wasn't even dry on our second batch of Stickies, and already we needed a much bigger batch. This time it would be $5,000 worth, to meet the demands of our expanding market. Money from the first order was trickling in, but we were already way behind the eight ball on our second order. And now there was this new demand. The vacation fund was empty. There wasn't enough money left in hard savings to pay for this third order of Stickies, and we were frittering away the checking account on things like food, clothing, and shelter. This was a crisis.

The only place left to look was the kids' education fund. We had an iron-clad rule that the education fund was sacred, not to be touched for any reason, ever, under any circumstances.

That afternoon I paid the printer for the new batch of Stickies out of the education fund.

And suddenly, that found money wasn't enough, either. Right on the heels of that order, we needed more Stickies—$10,000 worth this time. Sales were climbing sharply. All of this incredible growth had taken place in the space of about two and a half months, so there was never time for the incoming money to catch up with the outgoing money. I was becoming a Stickies addict, and my habit was getting awfully expensive to support.

Anyway, there we were, facing the same dilemma you'll be facing. Expenses and the pocketbook lag were eating us alive, and the fact that we had a winner on our hands only made the problem worse. We had to find a source of capital or we were going to drown in bittersweet juices of success.

The choices that face an inventor at a time like that are these: He can borrow from a bank. He can take on partners. He can go public and sell stock. He can start holding up liquor stores in his spare time.

We'll dispose of the last two alternatives fairly quickly. Moral considerations aside, you're not going to have enough time on your hands to be running around sticking up liquor stores. And as for selling stock and going public, it's remotely possible, but unlikely. By the time you go through all the red tape it takes to go public and by the time you've met all the requirements for a corporation that sells stock, it'll be way, way too late to do you any good. In any event, you can't make it to the big board. Or the little board, for that matter.

The next possibility is a bank loan. That's what I decided to try next when things got tough. I needed someone to loan me capital to keep my booming new business growing. Isn't that what bankers would have you believe they do all day? So I began groping around in my mind for a bank that seemed likely to listen to me. The best one I could come up with was a local bank that had once made me a loan—cosigned by a friend—for some other venture. I had paid the loan back promptly so my credit record was good with them. My name wouldn't be on their no-no list, anyway.

Better still, this bank had a reputation as a real go-go operation. They were supposed to be real swingers. Also, the branches of this bank were all autonomous. You could go down there and get an answer and get your money right away without waiting around for a stuffy bunch of cigar smokers at the central headquarters to make up their minds. I called the manager of the local branch—the guy I had dealt with on that previous loan—and told him I had a new product which was testing very well. Could I come talk to him about financing? I was running out of my own resources.

"Sure," he said. "Come on down."

There was hope. I really don't remember now why it was that I drove the station wagon down to the bank. But it was in the flower-covered station wagon that I made my grand entrance to the bank parking lot. I tooled in, found a parking place in the shadow of that vast expanse of chrome-and-glass architecture and stashed the Rickie Tickie Stickie-mobile.

The meeting was very pleasant. The manager expressed interest in the numbers I had worked up for my presentation. Then he wanted me to tell him a little more about exactly what this product was.

Just at this moment, his receptionist—bless her heart—strolled past the big window overlooking the parking lot. She took one look at the station wagon with all those flowers on it and came running the whole length of the bank, yelling to the manager. (I guess she was an old-time employee of the bank and didn't worry about formalities or the fact that this was the manager.) She came running up, screaming, "You've got to see this, quick!" She literally took him by the hand and dragged him over to the window and said, "Isn't that great! That's the greatest thing I've ever seen!"

He was sold. We got the loan, I'm convinced, strictly on the basis of the receptionist's excitement over a station wagon with the flowers on it. Unfortunately, we had to put up the house, the mountain lot, and just about everything we owned as collateral for that $10,000 loan. But I was sure that in just a short time the money would catch up with the expenses, and I'd be in a position to pay off the loan and fund the whole venture on my own. As it turned out, it was quite a while before that happened.

That brings us to a point about trying to borrow money from a bank to fund your idea. Banks are not very interested in ideas. They like collateral, thank you. The banker's human instincts told him that the flowers were really neat. His business instincts told him that my figures indicated a potential success. But his banker's instincts told him that the loan had to be backed by collateral, not by projections.

I've since been convinced that banks are not a very logical place to find funding for an idea. But I first learned it to my own satisfaction some months later when I went looking for another loan.

I was still suffering from pocketbook lag. But by now I had a good, solid sales curve to show. And I had a much more elaborate presentation. I asked my accountant for professional advice, and we put together a cash-flow projection for 1968 that should have warmed the heart of any knowledgeable investor or, as they prefer to be called, "financial partner." We had a full-bore presentation, complete with charts and graphs, full of impressive details. Then we spent days meeting with groups of bankers—swinging, go-anywhere, do-anything bankers; dour, pin-striped bankers; young bankers; old bankers—bankers who never could see the slightest connection between our little flower and their money.

One day was especially memorable. I had wangled an appointment with a bank known in the trade as bright, innovative, daring, and ready to take a flyer. Our kind of people. They would understand.

My part of the presentation was easy. I simply stood up and told the money men all about Rickie Tickie Stickies. It's like being asked to show pictures of your children. Anytime. Anyplace.

I showed samples. A sales brochure. Sales figures to date. Cost of goods ratios. Plans for national distribution and future new products. Advertising and promotion plans. All the stuff dreams are made of.

Thirty minutes of polite, if silent, attention was extended to my little dog-and-pony show. Then came time for Ken, my accountant, to do his thing. I've always been amused by the fact that great ideas are presented vertically, like a page from this book. Financial support for those ideas, however, always appears horizontally. The cash-flow charts ran off the right-hand side of the conference table.

Ken made all kinds of sense, it seemed to me. All we were asking for was a little credit line up to a million. "Up to one million what?" was the first question after a very long silence. I don't think I've ever heard a silence quite that long.

"Dollars," Ken replied. Then all the heads swiveled back to me. "What, again exactly, Mr. Krack, was that idea of yours?"

"A gaily colored plastic flower that sticks on things, sir," I shot back. And I added that the name was Kracke, as in *Business Week*.

More silence. To this day, I'm convinced that if our product had been a small black box that could be sold to the then-flourishing aerospace industry we would have had our credit line in five minutes flat. "Financial partners" don't understand black boxes any better that you or I. But, aerospace was a good thing.

If a black box flopped, you could at least feign confusion. But a gaily colored plastic flower? How do you make excuses to your lending committee when the plastic flower you backed with a million dollars turns out to be a loser?

We tried other banks. Same pitch. Same results. No bank was at all interested in lending that kind of money without collateral. I, of course, had no collateral left to put up. All I had was a wildly successful, profitable product, and the bank wasn't interested in that. It had all happened in the space of a few months, and banks like to watch a product perform for a good long time before they are willing to consider making a loan on it.

In the meantime, while you're watching your product perform, of course, and you're stuck in pocketbook lag, you may even be knocked off by some competitor and you are generally under-financed, overextended, and ready to turn turtle. By the time the bank has watched you long enough to be convinced that you're a success, they probably will have another problem. "You've peaked," they'll say. "Your idea was a nice novelty item, but we think that the demand is ready to slack off. We can't loan you any money on a product that will be slowly going downhill from now on."

And that will be that.

I suppose it's not really their fault. Bankers are trained to be cautious about lending money, to make sure that there's collateral backing any big loans. Their whole operation simply isn't geared to deal with quick success. Mediocrity they can fund. A product that plods along for years with no great spurts of growth can probably find backing. But that kind of product usually doesn't need bank funding. They don't get caught in pocketbook lag. A product that takes off like a skyrocket needs money to fuel that growth. But that's precisely the kind of product that banks are not willing to look at.

ALWAYS be FAIR

Let's leave banks, for the moment, and look at the other alternative I mentioned: partners. That's where I went, finally. The trouble with getting partners is that they want a share of the action in return for their investment in your product. That's not an illogical demand, but it does take some of the sweetness out of your success. Every profit dollar that rolls in has to be split up and doled out among *all* the partners. You're just one of them. You may be the major partner, but you still don't get all the profit from that marvelous idea of yours. It's a little galling. Especially when you're the one who had the idea, you're the one who's doing all the work, and you're the one who's taking the biggest risk.

Still, finding partners is a viable solution. I found two partners, when the pinch really hit. And instead of getting $120,000, which would have been enough to let me finally catch up, comfortably, with the pocketbook lag, I had to settle for much less. It was big money, of course, but it wasn't big enough to make things comfortable. Knock-offs were beginning to show up and skim the biggest part of the market I was trying to reach. I needed a lot more money and a lot more distribution to catch up with them. We had been through our negotiations with Chicago Printed String by then, so that avenue of escape was closed. It was up to us to run with the Rickie Tickie Stickies the best way we knew how. The investment by two new partners made it possible to do that—but barely.

And in getting those two partners, I had to give away a hunk of my company. *My company,* shared with strangers! (Well, they weren't strangers. As a matter of fact, they were friends, and the deal they asked for was not unreasonable, under the circumstances. I just hated to give away part of my dream.)

If you have to take on partners to keep your idea going, try to avoid giving away any more of the action than you absolutely have to. And if possible, keep open the option of buying them out at some later point in the life of your product. If your idea is going well, you may make enough to buy out your partner at the current value of his share of the company.

The earlier you can buy out your partner, the better, of course. Then, as the business continues to grow, you will recoup the cost of that buy-out. For example, at the point when you buy out your partner, his share of the company may be worth $50,000. He's doing well because he's been sharing in profits all along and maybe invested only $10,000 originally. Now, as your business continues to grow, the share you bought back for $50,000 might become worth $75,000 or more. So you wind up doing well. So does he.

Obviously, it's also always better to have "silent" partners—that is, partners who invest money but don't participate in the management of the company. If you're doing a good job of developing your idea, you don't need any more cooks in the broth. They're unlikely to be as well versed in your business as you are, anyway, so you're better off—and they're better off—with you running things. If they weren't satisfied with the way you were handling the business, they wouldn't have invested in the first place.

MOONLIGHT becomes THEM

There also is another kind of partner. This partner doesn't invest money, nor is he a "silent" partner. What this partner invests is his time and his expertise.

For instance, you'll need professional advertising and merchandising help to design your package, write copy for your collateral material, do your layouts for the collateral, design your letterheads, and all those other technical jobs I mentioned in the chapter on merchandising.

That can be expensive. A professional copywriter or art director can command anything from $100 to $200 an hour for his services if he or she is good. And the kind of work they do is slow. It takes a lot of time to write even a catalog sheet or a window banner. It's not just a matter of sitting down at a typewriter and pounding out a sentence or two. Likewise, in designing packages and layouts, the work goes slowly and carefully if it's done well. To do all the little jobs I mentioned in the merchandising chapter could easily cost you several thousand dollars' worth of creative time by these professionals.

You may be able to convince them to do the work on their own, with no money coming in, if you offer them a share of your business. Needless to say, that share ought to be proportional to the amount of work they do for you. But if your product does well, they could be much better off with a share of it than a regular fee. A job that might have brought one of them a thousand in cash might eventually bring in five or ten thousand. Gary Dahl is a writer by trade. So when he needed to have someone design the package for the Pet Rock, he offered an art director a percentage of the business as a fee for doing the work. The art director agreed and spent several days working out the design, which was excellent. By the time Gary's business grew enough to permit him to buy out that art director's share in the company, the art director got $50,000 cash for it. That's not bad wages for a design job. Especially when you've been making $15,000 a year!

Advertising agencies often frown on letting their people do freelance work like this, but almost every creative person I've ever met has moonlighted at one time or another. So, by getting to know someone at an advertising agency, you can get a line on who might be willing to work for you in developing your new product. Art schools, too, are a good source of technical help in this area. Either the instructors themselves or some of the advanced students in commercial art and design courses might be able to handle the job for you.

All of this is relatively insignificant, though, compared to another way you can work this same angle. Think, for a moment, about who is the most involved and most aware of your product other than yourself. Your suppliers, who else? The person who's printing your packages, the fellow who is manufacturing your product, all the people with whom you deal and to whom you have to pay money to get your product moving in the marketplace are potential sources of operating capital.

RUB a DUB, DUB

Give your printer a piece of the company in return for his services. It's probably a better deal for him than a straight cash operation in a couple of respects. First, as I said before, the value of his share of the company could become far greater than what he would have made in cash strictly as a supplier. Second, I don't know a printer anywhere who doesn't have downtime on his presses. So by slipping your work into those periods of idle time in his shop, he won't even be losing regular business. He can provide you with what you need at very little cost to himself. That's an attractive deal if he has some faith in your product. And if you've been dealing with him, paying your little bills promptly, and being straight with him, he's probably willing to listen to you. He's seen what has been happening, and he's probably picked up on some of your enthusiasm.

Let's face it, it's a lot easier to make one of your suppliers understand the problems you're trying to beat than it is to get a banker or investor to understand them. Your supplier has been living with those problems every day, the same as you have. He knows the score.

Since the manufacturing of your product and the printing of your packages and so forth represents such a large portion of your expense, you can turn a share of your business into the equivalent of an enormous amount of financial backing.

Of all the different ways you can choose to put some money behind your product, this one is far and away the best of them all. It not only gives you a lot more maneuvering room in your finances, but it gives you partners who are highly motivated to help you succeed. After all, the more money you make, the more money they make. Your success is their success.

16

Serendipity-Doo-Dah

After all the hard work is done, there's one more factor to consider: luck.

I'VE BEEN DOING A LOT OF TALKING ABOUT HARD WORK AND perseverance and doing your homework and keeping your nose to the grindstone and all that. It must sound like a lecture on the work ethic by now. You'll be happy to know that there's a more romantic factor involved in making a success of an invention. In all the talk about hard work, I've neglected to mention this one factor which is the most important influence on many a success story. It turns up too many times to be ignored.

The factor is luck.

Serendipity.

Chance, happenstance, coincidence, call it whatever you want. It's always waiting in the wings to make its fateful appearance on stage and change everything around. There are stories I know of that would never make it as pieces of fiction because nobody would believe them. Consider, for instance, the following.

THE CORNED BEEF HASH AFFAIR

I was in Chicago to make a presentation of my modular storage system— Neat Garage—to a manufacturer. This was going to be my big chance to finally get the idea turned into a product, and I was doing my best to talk them into giving me a nice royalty deal. I got out my portfolio with all the market estimates, sales projections, glowing descriptions, and full-color, 8″ × 10″ glossy photographs of my garage before and after Neat Garage. I talked. I waved my arms. I cajoled and explained and sold. Nothing happened. The company wasn't interested.

So I packed up my traveling minstrel show, whistled up my dogs and ponies, and headed back to the hotel where I was staying. By now it was well after lunchtime, and I was hungry, so I dumped the presentation portfolio in my room and went back downstairs to the dining room to see if I could still get something to eat.

By this time of the day the great, cavernous dining room—it must have been built to accommodate at least two hundred people—was echoing and almost empty. There were only two other people sitting down to a late lunch by the time I got there, and the hostess seated me next to one of them. I took a look at the menu and wasn't really inspired by anything I saw there. The guy next to me was working on something that looked promising, so I turned to him and asked, "What are you having?"

He looked up. "Corned beef hash," he said.

"Any good?"

"Try the omelet," he said, smiling.

I chuckled at that and pretty soon we had a conversation going. Good grief! His parents live in Palos Verdes, just about a mile from my house. We went through the whole "Small world, ain't it?" routine, and he asked me what I was doing in Chicago. I explained that I was trying to sell this marvelous idea for a modular storage system.

"Hey," he said. "That's kind of interesting. It sounds like something my company might be interested in. Tell me about it."

It turned out that he is a marketing executive for a major national building products company. Within minutes, we were back upstairs, and I had dragged my whole dog-and-pony show to his room where we were going over my portfolio of market data and pictures. He loved it. It was just the kind of thing he was looking for, so he set up an appointment for me to make my presentation to his new-products committee.

A few weeks later, I made that presentation to the committee. When we were done going over the details of the idea, the committee was as interested as my friend, the hash expert, had been.

If it hadn't been for a bad batch of corned beef hash at the hotel that day, I might never have gotten into that conversation. I would never have had my shot at that new-products committee. For all I know, the corned-beef hash might turn out to be a million-dollar stroke of good luck disguised as bad cooking. Come to think of it, I wonder what business the other guy in the dining room was in? Supposing I'd sat next to him? I guess I'll never know.

LUCK ÜBER ALLES

The Serendipity Goddess put a lot of time and effort into Rickie Tickie Stickies. The idea itself was born because of a stroke of coincidence. When we were driving down that L.A. freeway and came upon those decorated cars, we might not have paid too much attention if they had appeared one at a time. But fate sent all three of them past us in a row! That was too much to ignore.

The mental wheels started rolling, and the idea was born. I might never have become the Head Stickie if it hadn't been for those three specific cars showing up simultaneously at that particular spot on the freeway that day. Do you have any idea what the odds are against any three specific cars showing up together at a given spot on the freeway? I'll bet you'd be more likely to be dealt a natural royal flush in five-card poker.

The story of Polyoptics is partly a story of luck and partly a story of stick-to-itiveness. Polyoptics is a company that makes those plastic strands that conduct light the way a pipe conducts water. You can shine a light on one end of the strand, then weave the strand through all kinds of complicated curves, tie knots in it, and string it around corners. When you get to the other end of the strand, you've got light shining brightly wherever you choose to point it. It's nifty and a highly creative bit of technological work.

The company had been making the strands for a number of industrial and commercial applications, but there wasn't a whole lot of demand for it. Nobody was really geared to make use of the idea, and Polyoptic's sales were pretty slow. So, to keep from going stir crazy around the office and to try to find a new outlet for the product, the company's designers worked up a table lamp for the home. It used bunches of the Polyoptic fibers to conduct light out of a hidden central source. It looked like some sort of turned-on sea anemone or a wild bunch of electric hair on a lamp base, so it had quite a bit of novelty appeal. Some different versions of the new lamp were made and the company took them to the next trade show in San Francisco, where they hoped to break into the consumer market with this new application of their industrial product.

It began to look like a total flop when, after three days at the show, not a single sale had been made. People were staying away in droves. Finally, one store placed an order, just in time to keep Polyoptic from packing up and going home. Then another came in. Pretty soon the trickle had grown to a respectable flow of orders, getting the product off the ground. That broke the ice.

As the lamps hit the stores and people began buying them, orders grew larger and more frequent until Polyoptic had sold literally millions of dollars worth of lamps and was pulling in a solid profit. They even made money on the knock-offs, since they manufactured the crucial fibers. Better still, the sales of the lamps at retail began to put the strange new fibers out before the public eye. Industrial engineers and designers became aware of them in large numbers, and before long, a lot of new industrial applications were being developed. So, as a rub-off from the retail sales of the lamps, the company's industrial sales began to move out smartly, too. None of it, of course, would have happened if Polyoptic had packed up and left the disappointing San Francisco trade show before the first order came in to break the ice. Needless to add, that bit of serendipity wouldn't have happened if Polyoptic hadn't stuck to its guns after three days at the show. Sometimes Lady Luck needs a little help, too.

THE $89,000,000 GIG

I don't want to end this chapter leaving you feeling that somehow the Serendipity Goddess will bail you out with your idea. She won't get you out of the job of doing all that homework. You'll still have to do all the work, take all the right steps, and probably spend an uncomfortable amount of money if you are really serious about making your invention a success. But every now and then, Lady Luck *does* turn up where you least expect her, and she does the damnedest things.

We can't leave the Sea of Serendipity without my sharing with you a story that is unfolding concurrent to my finishing the last few chapters of this book.

It's serendipity with an overtone of doo-dah if I ever heard one. To start, we have to go back about fourteen years when a delightful young lady named Donna wandered into my office.

Actually wandered is not the right word. She had made an appointment through my patent attorney to review her dandy new-product idea. It was a beach mat. Although the idea was basically solid, I did not have any immediate contacts that could help her. I did, however, give her a copy of my book, wished her well, and told her to be sure to call as soon as the questions arose.

That chance meeting led to a bit of my psuedo-philosophy that supports my theory that if your primary intent in this life is to help others, not just make money, you will, in fact, get back a lot more than you give.

From that first meeting in 1986, and through at least ten more meetings over the ensuing fifteen years, I was never too busy to see (or talk by phone) with Donna. Why? It was the right thing to do. And, coincidentally, it paid off in spades.

As recently as at a business meeting in my home last Sunday, Donna introduced me to her business associate as her "mentor." That drove the point home for me. What I considered a simple relationship of helpful kindness, Donna had used as a life-changing experience. She had, over the years, transformed herself from a shy, introspective inventor into a business tycoon who casually laid an $89,000,000 gig in my lap.

So if you actually need a reason to be kind to others and share your experiences with them (for free, I might add), look no further than this little vignette.

Stop
THE Presses

17

"If you want praise, show it to your mother."
—DK 12/15/2000

AS FAR AS I WAS CONCERNED, AS OF LAST WEEK I WAS COMPLETELY finished with this book's first-draft manuscript.

Two weeks early, as a matter of fact. Now I'm late because of what (I hope) are a series of extraordinary events that happened to me during the last two weeks. They all came to a head on Friday, December 15, 2000, and I had to share them with you.

Here's the setup: On Saturday, December 2, 2000, my son David and I made an excellent, three-hour presentation to a very simpatico novelty gift and toy manufacturer in Portland. The company is named Hog Wild, and it is very much on the leading edge of clever gift and toy new-product concepts. My kind of folks. Having already made a get-acquainted pilgrimage to Hog Wild Headquarters on September 30, 2000, David and I were confident that the 35 new products we were presenting neatly fit into Hog Wild's product mix. Remember that those thirty-five new-product ideas were carefully selected from the 163 new-product ideas in my Creative Crap ideas files.

As an aside, the best thing that can happen to a new-product creator after a presentation is for the targeted company to request to keep one or two ideas for "further review." Hog Wild asked David and me to leave the whole batch of 35.

Now, I'm too professional to expect to hit thirty-five for thirty-five. However, on our way back to David's home, he and I agreed there were eight ideas that the Hogs had really warmed up to. Being realistic, we agreed that three or four would be moved ahead and no less than two would get to market. Our conclusion, of course, was that those two would become great hits and run for over one hundred years.

Hog Wild stated that they would be back to me by December 15, 2000. That, in itself, was good news. They were prepared to move swiftly.

We'll leave this part of the story at this point to review another new-product-line concept that I had working in parallel to the Hog Wild project. Entirely different product line. Entirely different manufacturer.

I think I mentioned that I owned the name Flower Children for the Toy Class 28. I also have created a line of copyrighted, cute dolls and doll accessories to match the line name. Add to that the fact that my entire Flower Power Homewares Retro Look is quietly moving into the classic stage of its life, and you can see why I was confident of acceptance when I sent the Flower Children Line off to a major toy manufacturer. I was even more confident when the vice president, New Product Development, of that major toy company said by phone, "I like the idea. And, if you don't object, I'd like one of our doll designers to take a whack at restyling the actual doll figures." I was elated. And, furthermore, in that same phone conversation, he asked if I'd give some thought to what advance against royalty I would want were they to move forward with my program.

I'm including here the response letter I sent to that VP at that major toy company that very afternoon:

Dear . . . . ,

Here's where I am on the royalty and the advance.
I believe a 7 percent royalty is in order since the name, Flower Children, is already trademark registered and the patent on my Flower Power Wheel has already issued. There is no blue sky here.

Here's how I arrived at the advance:
1. I'm guessing that you will do $2,000,000 in net, wholesale sales for year one.
2. 7 percent of $2,000,000 = $140,000 in expected royalties.
3. 50 percent of $140,000 = $70,000.
4. 50 percent of $70,000 = $35,000.

Voilà. The answer is . . . a $35,000 advance against a 7 percent royalty for the life of the product and all the attendant components.

I'll call you Monday afternoon to discuss this further.

Cheers,

Don Kracke

We'll leave this part of our story at this point and move on to yet another parallel universe I was exploring simultaneously to the Hog Wild and Flower Children projects.

This effort involved my registered trademark for the Home Air Freshener Class 5. When you coupled my Flower Power name with my trademarked flower designs and my design-patent-pending package designs, you ended up with an irresistible new packaging and merchandising concept for a major manufacturer of home air fresheners.

I found such a company, arranged an in-person meeting, and flew off to make my "on velvet" presentation. You know by now that setting modesty aside is easy for me. So, modesty aside, my presentation was magnificent. And, yes, the targeted manufacturer requested that I leave behind all my materials so the company's new-product development team could review my idea further. I was to call on December 15, 2000 to get their team's opinion. This part of the story will now be put on hold a moment so I can recap the situation:

* I've got thirty-five new-product ideas under review at Hog Wild, with the decision date being December 15, 2000

* I've got my entire Flower Children line of dolls and accessories at a major toy company, with the decision date being December 15, 2000

* I've got my entire Flower Power line of home-air-freshener products at a major home-air-freshener manufacturer, with the decision date being December 15, 2000

On December 15, 2000, all three companies passed on all of the forty-five SKUs I had presented. My out-of-pocket investment on these three projects to date? About $70,000. My return to date? Zip. Zero. Nada. Zilch.

Am I discouraged? No. But I really am pissed. However, that too will pass. And, by the first of the year, I'll be dusting off my hit lists for each of the categories and by the first of February, 2001, I'll be back on the road again trying to sell each and every one of those ideas.

And, does this story have an earth-shattering moral? Not exactly. But I did want you to meet reality face-to-face.

18

YEAH, BUT What HAVE You Done FOR Me Lately?

Trips and trops up and down the author's new-products memory lane.

TO MAKE SURE I HADN'T LOST MY WRITING TOUCH, I TOOK ADVANTAGE of a friend's offer to review this book's manuscript before I shipped it off to my publisher. Since my friend is a very successful, professional writer, it was an offer I could not refuse.

You can imagine my chagrin when my friend's summary note read as follows: "This is a pretty good book just as it stands. It could be greatly improved, were you to include a few examples of what you have created since the first book was published. Getting yourself into the twenty-first century would not be a bad idea."

The next thing that bastard will suggest is that I get a personal computer to replace my legal tablet and my trusty Pentel Markers. However, in thinking about it a bit further, my pal did have a valid point. So I tossed out my rather banal last chapter, dug into my Creative Crap file, and compiled a list of the things I have done lately.

There are two distinct categories for this list of items: (1) those that have made it to market; and (2) those that have not—yet. Where I was able, the hard-dollar investment I have made to date on each was noted. Where the information was readily available, any gross income from those having made it to market was also noted.

TO MARKET, TO MARKET

Here are some of my over 160 current ideas that I have licensed and that have gotten to market since my first book came out.

1. THE PORT/STARBOARD JACKET: This you know about from our third chapter together. To repeat: nine years on the market through the Land's End Catalog. *Gross Income: $47,500*

2. THE STUFFIN' FAMILY TEACHING DOLL: I did not personally create this little jewel. Sheila and Joel Kushell did. Think of me as the midwife who managed to lose $15,000 a year for three years in a row by trying to manufacture and market Little Lumpsie, my version of the doll. After I lost $45,000, I woke up to the fact that I would be better off licensing my little baby to a major toy company. That company was Fisher-Price. They sold 900,000 units. I made five cents a unit royalty. *Gross Income: $45,000*

3. POOL GLASSES: Think of fifteen plastic beverage glasses in a triangular serving tray (rack). The glasses are numbered one through fifteen. Seven are solid colors. Seven are striped. The Eighth glass is black. And, oh yes, there is an all white, cue glass included in the triangular shaped, full color display box. It's the perfect house, swimming pool, or game room party gift. At least at $19.95 a set retail it would have been. At the $34.95 Sharper Image was trying to sell the set for, I was snookered. To mix a metaphor (or two). And Epstein Design in Sausalito didn't cover the advance they gave me. *Gross Income: $6,000*

4. THE EXCEDRIN TABLET PAPERWEIGHT: Envision a five-inch diameter, one and one half pound plastic Excedrin Tablet complete with a debossed E on the top. Bristol-Myers bought the first 1,500 as salesmen's give-a-ways. The idea lasted about six years on the gift market. *Gross Income: $78,000*

5. LUGGAGE SPOTTERS: Twenty-four different brightly colored Hang Tag/ID Tags for luggage. All on a Floor Spinner Rack. A merchandising jewel. *Gross Loss: ($4,800)*

6. RICKIE TICKIE TWO: The company I sold Rickie Tickie Stickies to in 1972 went belly-up in 1974. *Gross Loss: ($60,000)*

7. RICKIE TICKIE THREE: With the rights having reverted to me by default, I tried to latch on to what I hoped would be a Flower Power Revival in 1988. That time around I also had my own knock-off line ready to go. Remember Put-Ons? Neither does anyone else. *Gross Loss: ($42,000)*

8. RICKIE TICKIE FOUR: I knew I couldn't miss by tying my flower decals to the thirty-year-anniversary celebrations of all things retro, which has been building steadily from 1997 right through last week. *Gross Income: $70,000*

9. RICKY TICKIE FOUR, PART TWO: I had already decided to manufacture and sell my flowers myself. (That credit-card discovery referred to in chapter 15 was a major factor in my doing-it-myself.) What I had not put in my marketing plan was my developing prostate cancer in 1998. Suffice it to say, that kind of news will really focus you on deciding what in the rest of your life is important. Stickies was certainly not one of those things, so I sold my restart-up company (on a $10,000 advance against a 10-percent royalty) to a major California-based gift company. *Gross Income: $10,000* *P.S. Fortunately, I won the cancer battle and am now back to work.*

10. OLD TIN CANISTERS: My then design partner, Rodger Johnson, created a great look originally designed for metal kitchen canisters. It proved to be the beginning of what became the Center for Homewares Design. I sold the idea to Ballonoff Home Products for Kitchen Canister Sets and, ultimately, over 180 different Old Tin SKUs made it to market. *Gross Income: $1,360,000*

11. DECORATIVE OUTDOOR VINYL WELCOME MATS: Royal Rubber did a yeoman's job of manufacturing and marketing Rodger's and my line of doormats for over ten years. The next time you wipe your feet on a plain, old doormat, wonder what else it could look like. *Gross Income: $32,000*

12. FANCY CANS GARBAGE CANS: Ever wonder who thunk up the idea of putting pretty pictures on twenty gallon, galvanized garbage cans? You're lookin' at him. And they're selling better today than when I created the concept back in 1975. *Gross Income: $67,000*

13. THE COUNTRY CRITTERS COOKIE JAR LINE: Treasure Craft licensed this very amusing line of cookie jars from us in 1997. The original series of farmyard critters was designed by Rodger Johnson's son, Gary. Really cute—if you like straw hat lids. My part: selling the idea to Treasure Craft. *Gross Income: $6,500*

14. THE MUSICAL CHAIRS DOLL LINE: My wife, Silvia, came up with this really clever idea. Picture a group of cute, hand-size dolls. Each sat in a chair which, when the doll was lifted up, would play a tune. One minor problem. The manufacturer to whom I licensed the idea forgot to make the dolls removable from the chairs. *Gross Income: $2,500*

15. THE CENTER FOR HOUSEWARES DESIGN, INC.: We're now slipping into the big time when it comes to the new-product business. Suffice it to say that The Center, which Rodger and I created and ran successfully for over twelve years, was the biggest success I have been associated with. Ever. The Center Story will probably be my next book. For the here and the now, realize that for each of twelve years, we created nearly two hundred designs for coordinated homewares.

    The Center itself had an exclusive membership of over fifty, noncompetitive housewares manufacturers. Each paid us an annual advance. Names like Anchor Hocking Glass and Rubbermaid were included. An interesting aside: At 12:52 P.M. PST today, Jill Parker from Decora (the company to whom Rubbermaid sold the Con-Tact brand) called to report that our royalty for September and October 2000 came to a bit over $1,700. That would just tip the royalty income on the shelf liner license to a bit over $1,300,000. (Those 2.5 percents do add up.) Major Center design themes I'm sure you would recognize include our Greenhouse, Orchard, and Blue Belle patterns. *Gross Income: $26,000,000*

16. THE HOMESPUN HOLIDAY COORDINATED CHRISTMAS COLLECTION: Target Stores selected The Center (after a fifty-design

bake-off) to produce the first coordinated Christmas design issue offered at the mass-market level. We worked with twenty-six different manufacturers. It was worth it. *Gross Income: $187,000*

17. KISS BELOW THE MISTLETOE: My number-one son, David, would rather be an inventor than the attorney he actually is. He is, however, the perfect candidate for the UCLA Extension Course I created and taught up until that pesky, old cancer visit. The course was called "How To Get Your Idea or Invention to Market (Without Giving up Your Day Job)." Every few weeks, David calls with another great, new idea. Six months ago he suggested that mistletoe on a belt buckle would make an excellent Christmas Novelty Gift. I didn't like suggesting to any of the gift companies I knew that they go into the belt business. Way too complicated. On the other hand, were you to attach mistletoe to a handy-dandy clip, one could attach it anywhere on one's bod. Laid Back Industries in Oklahoma City liked the idea to the point that it promoted it as a hot, new Christmas item for the year 2000. Incidentally, when I thought of the clip application, I became David's fifty-fifty partner. *Gross Income: $452.22*

18. THE YANKEE DOODLES COMIC STRIP: Two of my former partners in the creative business, Fred Martin and Ben Templeton, had created an excellent comic strip staged in the Revolutionary War period of our country's history. The only problem: None of the six national syndicates that were shown the property over the previous six years liked it enough to give it a try. I came across their presentation quite by accident one day in our storeroom. (Serendipity Doo-Dah.) "If I were to sign up a syndicate, could I become a one-third partner in the property?" I asked. "Of course," they replied. The next day my friend (and three-time Pulitzer Prize–winning editorial cartoonist) Paul Conrad stopped by my office, reviewed the strip, and liked it. The following day, he and I were in the office of *The Los Angeles Times* Syndicate's director. He, too, liked Yankee Doodles. The fact that our country's bicentennial celebration was only two years away helped a lot. Six years prior had been too soon. The next day we signed a contract. It ran three years. *Gross Income: $72,000*

19. THE YANKEE DOODLE LUNCH BOX: This little dandy (Get it? Yankee Doodle Dandy?) taught me a hugely valuable lesson. Licensing one's creative property is one helluva lot easier than creat-

ing the property being licensed. If you were paying attention to the comic strip income total and divided it by the three years the strip ran, you'd quickly see that three guys divided up $24,000 a year. That's for 365 original, funny strips per year. Our license to the Thermos Company generated $27,000 in one year for the five drawings we lifted from the strips. *Gross Income: $27,000*

19. THE 200 YEARS AGO TODAY DAILY COMIC PANEL: Remember my advice about expanding upon your success? I followed that advice. Ben, Fred, and I created and I sold to the *Chicago Sun Times* Syndicate a feature called *200 Years Ago Today.* It ran three years, too. *Gross Income: I can't remember.*

20. *THE INTERNATIONAL HANDBOOK OF JOCKSTRAPS* BOOK: None of my publisher friends would even consider taking this property to market. Undeterred, I self-published, sold the book only in gift stores, positioned it as "the perfect gift for your favorite jock," and sold 210,00 copies over a nine year period. So there! *Gross Income: $315,000*

21. THE *HOW TO SUCCEED IN BUSINESS WITHOUT KISSING ASS* BOOK: Again, every publisher friend I knew rejected this wonderful piece. The jerks! Undeterred, my writing partner, Lee Whipple, and I self-published. So far we have sold about 100 copies and we have given away about 900. *Gross Loss: ($8,000)*

22. THE *SUCCESS STATE OF MIND* BOOK: Lee Whipple spent a lot of time backing out "the attitude" we had purposely built into our *Kiss Ass* book. I don't have a clue as to how many he printed or sold. *Gross Income: ?*

23. PARENT PROTEST POSTERS: My long-time friend, Jerry Cooper, dropped by the office several years ago with layouts on a line of small posters he wanted to license to me. I liked them, the gift trade liked them. *Gross Income: $140,000*

24. GEMMA'S PEOPLE: A famous artist friend of mine, Gemma Taccogna, created a series of fanciful, life-size figures. The kicker: They were silk-screened on plain, brown, corrugated cardboard, and die-cut to approximate life-like, el cheapo room decorations. Great product. Rotten response from the public. *Gross Loss: ($6,250)*

25. THE AGING OF AMERICA COLLECTION: In order to get myself noticed as a fine artist, I created this series of three-dimensional art objects: A life-size Bidet on Rockers. A box of Argo Condom Starch. A Surfboard with a chair bolted to it. An *Eighty-Seven Ways to Prepare Cat Food* cookbook, and a 4′ × 7′ eye chart. To date, not one has sold, even though I published an expensive catalog and featured them prominently in my art gallery for a year. *Gross Loss: ($12,600)*

25. THE REALLY RETRO RICKIE TICKIE STICKIES HOUSEWARES COLLECTION: Just before the cancer thing, I licensed this line to Wilton Industries in Chicago. I'm convinced that had I not had to drop out for two years, I could have pushed the Wilton folks to a point where the line would have been more successful. It always takes a "pusher" on any new product. Even from the outside. I did get a $30,000 advance against royalty. There's still about $2,500 to be worked off the advance. *Gross Income: $30,000*

27. THE BAG LADY PAPER SCULPTURE: Yet another grand attempt to make a name for myself in the fine art biz. No one cares. *Gross Loss: ($3,600)*

28. THE SLINGSHOT RACE CAR TOY: Sears retained me to turn a bunch of aluminum tubing, some garbage can tote wheels, and some canvas into a toy. I did. 24,000 sold out at $24.95 in a one-weekend test. I was paid a flat fee. Not a good idea, in retrospect. *Gross Income: $5,000*

29. THE FATHER CHRISTMAS GIFTWRAP AND TOTE BAG LINE: Paperfest out of Miami retained Rodger and me to create a line of Christmas merchandise. We did our job nicely, nicely. Paperfest could not get the production costs far enough down to make an adequate profit so they dumped the line. *Gross Income: $39,000*

30. DINOTROLLS: Rodger's and my friend, Sandy Walkes, created a new toy product we all thought was a boffo concept. We did the design. Significant sales were just starting when that big, purple bastard came along, sat on our DinoTrolls and squashed our sales. *Gross Loss: ($3,500)*

31. BAGZILLA, MONSTEROUSLY STRONG TRASH BAGS: I was retained by Sears to create a name and graphic signature for a superior line of trash bags they were going to introduce. Brute Strength was the leading contender until one of my then partners, Art

Kushell, came up with the Bagzilla name. Ben and Fred added the words and graphics and the project won a Sears Tower Award as one of the best new product introductions by Sears that particular year. *Gross Income: $35,000*

HOW LONG are YOUR ARMS?

Incidentally, before we get into the ideas I've worked on lately that have not yet made it to market, I hope you noticed the recurring theme of my co-venturing with a bunch of creative people. There's no way any of us could hack out this much new product on our own. Keep this in mind when your ideas start coming faster than you can handle them. Don't have as many partners as I did. Rather, have more "Arms-Length" co-ventures with you always retaining the controlling vote.

That said, let's get on to fantasyland. Here's a batch of fully developed, new product ideas that I have not been able to connect to the marketplace. Yet.

1. THE HALLOWEENIES CANDY LINE: I've been selling these little candy treats both in and out of the state of California for so many years that the trademark has registered. And all the line, including Weenies, is also covered by copyrights. All I need now is a licensee. *My out-of-pocket costs to date: $4,175*

2. FLOWER POWER HOME AIR FRESHENERS: The good news is that after just sitting around for a couple years doing nothing, this category suddenly seems ready for my line of products. The name is registered and I have design patents applied for on the packaging. *My out-of-pocket costs to date: $2,675*

3. FLOWER POWER CAR AIR FRESHENERS: I've been turned down cold by four different manufacturers in the category. I don't have a clue as to why. My flowers, in my opinion, are a lot more fun than a crummy pine tree. *My out-of-pocket costs to date: $1,350*

4. THE THREE NOVELTY LAMPS LINE: I have a matching set of three design patents. The most recent issued last week. Picture a lamp in the shape of an eighteen-inch push pin. Another is a twenty-inch screw. The last is a twenty-four inch light bulb that just sits there looking sterling. The three different turn-downs to date were all for the same reason, "Nice idea. Does not fit our marketing plans at the moment." *My out-of-pocket costs to date: $4,375*

5. THE GAMEPLAY BOARD GAME AND PUZZLES LINE: Two partners on this program, Mark Mugrage and Dan Hollembe are as disappointed as I that after over four years of trying, I could not get any game maker interested. Until yesterday. *My out-of-pocket costs to date: $3,485*
P.S. Mark and Dan must have over 500 hours of development time in this project as well.

6. THE *IT SEEMED LIKE A GOOD IDEA AT THE TIME* BOOK: This coffee table book for inventors is completely edited, every one of the 166 pages have been laid out through the photocomp stage, type has been set and no publisher in America has shown any interest in this tribute to my ego. Oh, did I mention the $80,000 (at cost) batch of full-color photos I am willing to toss in for free? *My out-of-pocket costs to date: $6,750*

7. DISEÑOS EN FLORES LEASING PROGRAM: Way too complicated to describe here. Suffice it to say it's a biggie. No takers since I haven't presented it to anyone. *My out-of-pocket costs to date: $9,400*

8. DANDY CANS: This fully patented line of decorator garbage cans is probably one of my finest new new-product concepts ever. I believed that in 1970, 1980, 1990, and the day before yesterday. Rubbermaid liked the concept but passed because they could only see about $20,000,000 a year at wholesale. You figure out what 5 percent of $20,000,000 comes to. *My out-of-pocket costs to date: $62,000*

9. THE COMPANY STORE RETAIL GIFT SHOPS FRANCHISE PROGRAM: The nicest thing I can say about this concept is that the prototype store (which I designed) was beautiful. The worst thing I can say is that we lost $8,700 in the first and only year of business. *My out-of-pocket costs to date: $14,600*

10. THE $100 BILL POSTER: This oversized photo blow-up of the new, $100 bill has a UPC code printed on it. It is one of my LA-LAist art-exhibit pieces. Prospects: Dim. *My out-of-pocket costs to date: $140*

11. THE "CLASSIC" FLOWER LINE: I just discovered where that extra $100,000 I thought I had in the bank went. It went to hell. Directly to hell. It did not pass Go. What I'm referring to is my abortive attempt to resurrect the old Rickie Tickie Stickies as a classic graphic icon. Actually, had that pesky, old cancer not reared its ugly head,

I probably would have had my money back by now. And, no, I've not given up. Not by a long shot. With nine different licensees on board, I've already gotten back about $72,000 of the $98,115 I've spent. Breakeven is now an acceptable mark. *My out-of-pocket costs to date: $26,115*

12. THE SNOWMAN CHRISTMAS WATER GLOBE AND CHRISTMAS GREETING CARD AND POSTER DESIGN: Just envision your basic, classic, cuter-than-a-bug's-ear little snowman out in the field surrounded by his forest friends. It's snowing. Gently. The only thing wrong with this Christmas Classic is that I put the carrot in the wrong place. *My out-of-pocket costs to date: $385*

13. THE GIANT COASTERS AND MATCHING MUG SETS: I wondered one day why coffee mug coasters were so small. "Why not have big, pretty, colorful, cut-out coasters to add a bit of festivity to a coffee clatch?" I asked no one in particular. Since no one answered, I went ahead and comped up the idea. Looks good. Pitched Zak Designs in Spokane and Hog Wild in Portland. So far—nuthin'. *My out-of-pocket costs to date: $695*

14. THE DECORATIVE FLASHLIGHT LINE: Envision, if you will, a line of twelve different flashlights that look like pepper mills, bud vases, candles, and the like. Anything but flashlights. My logic: When the lights go out, who can remember what drawer the flashlight is in? But, if you had a flashlight that was attractive enough to leave out on a shelf all the time . . . *My out-of-pocket costs to date: $1,850*

15. THE CLOTHESPIN KIDS DOLL LINE: Remember Sondra Cutcliff and the Christmas Ornament story? Well, combine that with my "Line Extension" theory from chapter 2, and you will know immediately where this concept is going. Since Sondra made the prototypes, I'm in this project with nothing but time. *My out-of-pocket costs to date: $0.00*

16. THE PLANT PARASOLS LINE: Ever watch tiny seedlings wilt in the hot summer sun? So did I. Why not make little umbrellas you could pop up when the sun was too hot for your plantlings? *My out-of-pocket costs to date: $650*

17. THE PLANT-A-LOONS LINE: This is the product line I referred to when I described my son David's misadventure at the Patent

Repository in Salem, Oregon. Regardless of the screw-up in the patent search, nothing should stop me from contacting the current patent holder and suggesting we do a joint-venture product. *My out-of-pocket costs to date: $3,450*

18. FLOWER ARRANGING JUST FOR KIDS: My wife Silvia is an expert silk flower arranger. We decided to create a paint-by-numbers-type craft kit that taught young people the joys of flower arranging.

   The president of the company to whom I licensed the line died just after we had signed the licensing agreement. The survivors running the company abandoned the idea. The rights reverted to me. I pitched the idea to a gift company last weekend with the initial response being positive. I'll let you know in my next book how we did. *My out-of-pocket costs to date: $6,750*

19. CANDLE LITES: If you were to take the labels off all the "Lite" beverages there are on the market and wrap them around glass patio candleholders, you'd see in a minute what I had in mind. *My out-of-pocket costs to date: $1,650*

20. THE PORT/STARBOARD WINE LINE: A gentleman whose name I cannot for the life of me remember, came to me with this very clever extension of my Port/Starboard Jacket concept. We comped it up and I presented the idea in all its glory to a major California winery. Those folks were way too serious about their wines. However, I'll keep trying. *My out-of-pocket costs to date: $1,625*

21. WALTER, THE ADULT CEREAL LINE: Watch for my next book called *It Seemed Like a Good Idea at the Time*. Walter, Wally, and Walt are given a whole chapter. *My out-of-pocket costs to date: $3,600*

22. THE CHEAPSKATE LINE OF KIDS CASUAL SHOES: To a very high rubber sole of a tennis shoe you add roller skate wheels in bas relief. They don't turn. The patent is pending, and the toddler version is, of course, called "Training Wheels." *My out-of-pocket costs to date: $1,250*

23. THE NEW TOTE BAG LINE: Don't forget that any successful graphic design will probably recycle successfully every fifteen years or so. My soon-to-be recycling of some very clever tote bag designs I did in 1970 will prove that point once again. *My out-of-pocket costs to date: $3,900*

23. THE CLASSIC FLOWER LINE OF KIDS WATCHES: If this sparkling line is so great, why have six major U.S. novelty watch marketers passed on it so far? *My out-of-pocket costs to date: $900*

24. THE CHEER OF FLYING BOOK: I really should finish this project. *My out-of-pocket costs to date: $2,400*

25. THE FUNNY FARM LINE OF PLANT-FOOD PACKAGING: Rodger Johnson and I created this snappy little line, and I have pitched it to Black Magic so far. They pitched it back. *My out-of-pocket costs to date: $2,200*

26. THE ZODIAC TOY LINE: As it will be with my newly rediscovered Tote Bag Line, so will it be with my once-successful Zodiac Design Line. Recycling old art is one of the only advantages I know of getting older. *My out-of-pocket costs to date: $0.00*

27. THE SGT. MAJEWSKI ANTI FATIGUE MATS: Ever wonder why they call them "fatigue mats"? So did the good Sergeant. *My out-of-pocket costs to date: $3,500*

28. THE FLASHLIGHT CAR: My friend (and fellow inventor), Darold Cummings, came to me one day with a great idea that made a toy out of a standard handled flashlight. I paid $750 to have a beautiful, 3D model made, and there it sits . . . *My out-of-pocket costs to date: $750*

29. THE TRYKE: Darold also created (and holds the patent on) a tricycle in which he put the two rear wheels up front. There is no reasonable explanation as to why the four major bike makers in America have crapped on this splendid concept. *My out-of-pocket costs to date: $9,650*

So there you have it. That's what I've done lately. I've included all this extraneous information for a very real reason. I want you to know that this is the work of a person with a passion for the idea business.

Unless you, too, have such a passion, perhaps you should consider another line of work. If, on the other hand, this list of silliness makes consummate sense to you, you're my kind of folk. Get to it. And good luck.

THE Inventor's Checklist

HERE IS A CHRONOLOGICAL LISTING OF THE STEPS YOU SHOULD FOLLOW when an idea pops into your head. At several points in the checklist, you will be confronted with Go/No-Go decisions. These are the points at which this checklist could save big money. Your money.

Is there a guarantee that if you follow this outline faithfully to its logical conclusion, you will make money with your idea?

No!

However, if you follow this outline, are you guaranteed that your idea will have been given the best possible chance of success?

Yes!

☐1. Have the idea.

☐2. Start a diary listing every action you take along the way, and have it notarized periodically.

☐3. Evaluate your idea for need, desirability, and profitability.

☐4. Do some testing of the idea, either on your own or by hiring professionals to back up your own evaluation.

GO/NO GO

If, at this point, your idea can't cut it, it's a good time to stop and get another idea. If you're still convinced you've got another Pet Rock, keep going regardless.

☐5. Work out the physical design by making a model, prototype, or sample run.

☐6. Determine your cost of goods, including materials, labor, packaging, plus one-time costs such as package design, production art, and special dies or molds.

☐7. Select the primary method of distribution for your product.

☐8. Set up your own "knock-off plan" for selling the same product under different names through different distribution systems.

☐9. Plug the costs of each system of distribution you'll be using into your cost calculations.

☐10. Arrive at your wholesale and retail price.

GO/NO GO

If everyone along the line of distribution can't make the kind of money he normally expects, this is the next place to consider stopping. Or at least weeding out the distribution channels that don't seem to make sense for your product.

☐11. Apply for a patent, copyright, or (if applicable) trademark registration.

GO/NO GO

If it turns out to be a protectable idea, move on. If not, but you think you can get massive distribution quickly, move on. If it is not protectable and you don't think you can saturate the market before the idea thieves notice you . . . do some soul searching before deciding to continue.

☐12. Watch out for the rip-offs. They get their mailing lists from the patent applications, among other places, and they'll be getting in touch with you.

☐13. Prepare your presentation. You'll certainly need it for selling your idea to a manufacturer. You may also want it to use in the search for financial backing.

☐14. Choose the manufacturers to whom you will try to sell your idea.

☐15. Find out who is in charge of new-product ideas at those companies and call or write them. Don't be discouraged if that person tries to talk you out of even trying.

☐16. If all else fails, say that your idea is already in the mail.

☐17. Then run like hell to the post office.

☐18. Request a copy of the target company's idea-submission form and sign it. It marks you as a professional.

☐19. Note in your "idea diary" the names of all the people you contact and the companies they work for.

☐20. Plan what demands you will make, what deals you will accept in your negotiations with a manufacturer.

☐21. Negotiate. If they agree to buy your idea, ask *them* to prepare the contract and let *your* attorney review it. That saves you some money, and if they're interested in your idea, they won't mind.

GO/NO GO

If nobody is interested in your idea after your whole sales pitch is given, it's time to consider abandoning the idea and getting a new one. On the other hand, if you're still convinced you have a viable concept, take the next big plunge.

☐22. Set the wheels in motion for a limited, local test-marketing sales effort.

☐23. Try to use "leverage" by getting a major distributor on your side.

GO/NO GO

At this point, money definitely rears its ugly head. If you can't finance the local test marketing and if you can't get financial backing or leverage on your side, bail out. If you can lubricate the financial wheels somehow, keep moving!

☐24. Produce a prototype run of your product—enough for a reasonable number of reorders.

☐25. Produce your merchandising tools.

☐26. Start selling, either on your own or with reps.

☐27. Record your sales results and the amount of play your publicity gets. Consider using a clipping service to keep track of the "ink" you get.

☐28. Wait six months. (It takes that long to get a good idea of sales performance, unless your idea takes off like a rocket.)

GO/NO GO

If the public reaction to your publicity and your product is a giant yawn, perhaps this is a good time to think of a new idea. If, however, your product is performing respectably, move on.

☐29. Try to sell your idea to a manufacturer again, using your test-market performance as a major selling point in your updated presentation. Go back to some of the people who turned you down before, as well as some new people.

☐30. Revise your notion of the deal you'd accept. With hard sales performance, you can afford to ask for more.

☐31. If some manufacturer buys your idea, have your attorney prepare the contract and let their attorney review it. You'll be putting your feet to the fire this time, so your man ought to write out the terms. Just don't be too greedy.

GO/NO GO

If you still can't find a buyer for your idea, you have a really tough decision to make: Should you take the plunge and roll out nationally on your own? Or is it time to pack up your grits and gravy and shuffle back to the drawing board?

☐32. If you're still game, analyze the costs of a full-scale national roll-out, including costs of staffing up, a national sales program, and market support.

GO/NO GO

If you can't afford it, the game is over. Deal out a new hand and go back to square one. If you can afford it or if you can get leverage or financial backing to make it possible, get back into the fray!

☐33. Rewrite your marketing plan, including one-, two-, and three-year sales projections. Work out cash-flow charts. Reanalyze your distribution channels.

☐34. Roll out.

The ultimate Go/No Go is now riding on public reaction to your product. If it's a Go and you succeed—smile a lot. You've earned it. And get to work on your next idea.

If it's a No Go, and your idea fails—cry a little. You've earned it. And get to work on your next idea.

Good luck! From those of us who have succeeded. From those of us who have failed. From all of the above.

A Typical Marketing Plan

THE FOLLOWING IS AN OUTLINE OF THE MARKETING PLAN USED BY A major national retail chain for any new product in its line. It asks the questions any manufacturer or marketer will want to know about *your* product when they're considering licensing it. You ought to know the answers to all of these questions if you plan to make a sale.

I. How does the idea fit into the existing product category?
(In other words, why is there an opening for such a product in the line, how will consumers perceive it, how does it find its place in the existing line?)

II. Where does the new product fit into the existing channels of distribution?
A. Catalog/Retail—The ones you and I know about already.
B. Commercial—Most of us aren't aware of it, but most major retailers sell to commercial users of products, too.
C. Industrial—Many products are bought and used by industry. One example is a line of hand tools—wrenches, hammers, chisels and the like.
D. Service Centers—For every group of retail outlets, there is a service center to maintain and repair what it sells. These centers are also retail outlets for products, including replacement parts.
E. Export—Yes, everyone sells overseas these days, too.

III. Sales and Profit Objectives.
A. How many units do we plan to sell? For how much?

IV. Marketing Strategy.
    A. Positioning—Top of the line? Cheapest? Most versatile?
    B. Pricing Structure—Are there deluxe and standard models? At what prices do these sell?
    C. Geography—If it makes a difference where the product is sold, where *will* it be sold? A new kind of snow shovel might be a gas in Minneapolis and a bomb in Phoenix, for instance.
    D. Competitive Rebound—Will the competition react strongly? How soon will they come out with a similar product? How similar? Will it be lower or higher priced, and how can we counter their move? In short, how long can we expect to have an exclusive on the new product, and what should we expect to have to do when the competition moves in?

V. Tactics.
    A. Advertising and Promotion Expenditures—How much?
    B. Nailing Down of All Details—When will the product be introduced? Will it roll out on a geographic basis or on a calendar basis or what? Where should the product be warehoused and in what quantities? A million different questions.
    C. Preparation of All Capital Expenditures—How much will it cost to get this idea rolling, and when will we have to start spending how much?
    D. Measure Feedback—What's the expected feedback?
        1. From customers.
        2. From sales people.
        3. From service people.
        4. From the credit department.
    E. Follow-Up Positioning
    Can the new product lead to a couple of new products? E.g., can the new paint scraper in the hardware department find a place in the sporting goods department as a boat-maintenance tool? Will the new chafing dish lead to sales of chafing dish fuel? Will the new back-yard barbecue lead to an interest in a new line of outdoor entertaining equipment?

VI. Measurement (all products are measured on market penetration).

A. On its own, as a single item in a total universe of retail items.

B. As part of a line—sporting goods, for instance.

C. Catalog sales response can be used to predict retail sales.

1. The catalog is usually the first place you see a new item at this major retailer.

2. Catalogs are also considered as part of the advertising media mix, along with radio, newspaper, television, and the like. The catalog not only sells a product, but it actually advertises it, too. It has been found that people who have a catalog spend two-and-a-half times as much in their local retail store as they spend in the catalog. (For whatever it's worth, credit card customers also spend 50 percent more in cash than they charge, so our major retailer can use bill stuffers as an advertising medium and as a means of testing response to a new line.)

Resources FOR Inventors

BOOKS

Gregory Battersby and Charles Grimes. *The Toy and Game Inventors Guide.* Stamford, Conn.: Kent Press, 1996.

Sonny Black and Gary Ahlert. *Selling Your Idea or Invention.* New York: Birch Lane Press, 1994.

Tom Bellavance and Roger Bellavance. *Inventing Made Easy.* Moosup, Conn.: Quiet Corner Press, 1999.

Don Debelak. *Entrepreneur Magazine: Bring Your Product to Market.* New York: John Wiley & Sons, 1997.

Don Debelak. *Think Big—Nine Ways to Make Millions from Your Ideas.* New York: Entrepreneur Media, 2001.

Bob Dematteis. *From Patent to Profit: Secrets and Strategies for the Successful Inventor.* New York: Avery, 1999.

Ronald Docie. *The Inventor's Bible.* Athens, Ohio: Docie Marketing, 2001.

Stephen Elias. *Patent, Copyright, Trademarks.* Berkeley: Nolo Press, 2001.

Stephen Elias and Kate McGrath. *Trademark: Legal Care for Your Business and Product.* Berkeley: Nolo Press, 2001.

Robert Gold. *The Entrepreneurial Inventor's Guide.* Upper Saddle River, N. J.: Prentice Hall, 1994.

Stanley I. Mason, Jr. *Inventing Small Products for Big Profits.* Menlo Park, Calif.: Crisp Publications, 1997.

Robert Merrick. *Stand Alone Inventor.* Sunnyvale, Calif.: Lee Press, 1998.

Thomas E. Mosley. *Marketing Your Invention.* Chicago: Upstart Publishing, 1997.

Robert Park. *Inventors Handbook.* White Hall, Va.: Betterway, 1986.

David Pressman and Fred Grissom. *The Inventor's Notebook.* Berkeley: Nolo Press, 2000.

David Pressman. *Patent It Yourself.* Berkeley: Nolo Press, 2000.

David Pressman. *How to Make a Patent Drawing Yourself.* Berkeley: Nolo Press, 1999.

David Pressman and Richard Stim. *Patents for Beginners.* Berkeley: Nolo Press, 2000.

Harvey Reece. *How to License Your Million Dollar Idea.* New York: John Wiley & Sons, 1993.

Judy Ryder. *Great Ideas into a Great Success.* Lawrenceville, N.J.: Peterson Guides, 1995.

Richard Stim. *License Your Invention.* Berkeley: Nolo Press, 1998.

James White. *Will It Sell? How to Determine If Your Invention is Profitably Marketable (Before Wasting Money on a Patent).* James E. White & Associates, 2000.

OUT-OF-PRINT BOOKS

Look for these in your local library, at used bookstores, or at Amazon.com, which has a great used-book merchant connection.

Ronald Docie. *Royalties in Your Future: An Inventor's Guide to Marketing and Licensing Inventions, Patents, and Technology.* Athens, Ohio: Docie Marketing, 1998.

Suzanne Duret. *Inventing for Wealth.* [City]: RGD Publishing, 1993.

Reece Franklin. *Inventors Marketing Handbook: A Complete Guide to Selling and Promoting Your Invention.* [City]: AAJ, 1989.

Gary S. Lynn. *From Concept to Market.* New York: John Wiley & Sons, 1989.

Deborah Neville and Robert Coleman. *The Great American Idea Book.* New York: W. W. Norton, 1995.

WEB SITES

Entrepreneurial Edge Online
www.edgeonline.com

Entrepreneurial Magazine
www.entrepreneurmag.com

Fast Company Magazine
www.fastcompany.com

Idea Café
www.ideacafe.com

Inc. Magazine
www.inc.com

National Congress of
Inventor Organizations
www.inventionconvention.com/ncio

SCORE: Service Corps of
Retired Executives
www.score.org

United States Patent and Trademark
Office (USPTO)
www.uspto.gov

USPTO Independent Inventor
Resources Site
www.uspto.gove/web/offices/com/iip/index.htm

United States Small Business
Administration
www.sbaonline.sba.gov

Index

ABOUT THE AUTHOR

Don Kracke, inventor and best-selling author, holds 11 patents, 86 trademarks, and 1,206 copyrights. He has created 2,300 different new products that have generated in excess of one billion dollars in retail sales and has served as the creative consultant to Campbell's Soups, Coca-Cola, General Mills, K-Mart, Mattel, Sears Roebuck, Target and Wal-Mart. He has made guest appearances as a new product development expert on thirty-four different television shows, including *Nightline,* the *Oprah Winfrey Show,* the *CBS Evening News,* and *The Regis Philbin Show.*
He is the winner of 217 awards for advertising, merchandising, and marketing design excellence. The author of six books and cowriter of two nationally syndicated comic strips, he lives in Lake Arrowhead, California.

ALSO BY DON KRACKE

How to Succeed in Business Without Kissing Ass, with Lee Whipple

Success State of Mind, with Lee Whipple

The International Handbook of Jockstraps, with Fred Martin and Ben Templeton

Some Straight Talk for the Average Guy About Prostate Cancer

BOOKS FROM ALLWORTH PRESS

The Entrepreneurial Age: Awakening the Spirit of Enterprise in People, Companies, and Countries by Larry C. Farrell (hardcover, 6¼ × 9¼, 352 pages, $24.95)

The Soul of the New Consumer: The Attitudes, Behaviors, and Preferences of E-Customers by Laurie Windham and Ken Orton (hardcover, 6¼ × 9¼, 320 pages, $24.95)

Dead Ahead: The Web Dilemma and the New Rules of Business by Laurie Windham and Jon Samsel (hardcover, 6¼ × 9¼, 256 pages, $24.95)

Emotional Branding: The New Paradigm for Connecting Brands to People by Marc Gobé (hardcover, 6¼ × 9¼, 352 pages, $24.95)

The Copyright Guide: A Friendly Guide to Protecting and Profiting from Copyrights, Revised Edition by Lee Wilson (paperback, 6 × 9, 192 pages, $19.95)

The Trademark Guide: A Friendly Guide to Protecting and Profiting from Trademarks by Lee Wilson (paperback, 6 × 9, 192 pages $18.95)

The Patent Guide: A Friendly Guide to Protecting and Profiting from Patents by Carl W. Battle (paperback, 6 × 9, 224 pages, $18.95)

The Advertising Law Guide: A Friendly Guide for Everyone in Advertising by Lee Wilson (paperback, 6 × 9, 208 pages, $19.95)

Legal Forms for Everyone, Fourth Edition by Carl W. Battle (paperback, includes CD-ROM, 8½ × 11, 224 pages, $24.95)

The Money Mentor: A Tale of Finding Financial Freedom by Tad Crawford (paperback, 6 × 9, 272 pages, $14.95)

Money Secrets of the Rich and Famous by Michael Reynard (hardcover, 6¼ × 9¼, 256 pages, $24.95)

Old Money: The Mythology of Wealth in America, Expanded Edition by Nelson W. Aldrich, Jr. (paperback, 6 × 9, 340 pages, $16.95)

Estate Planning and Administration: How to Maximize Assets and Protect Loved Ones by Edmund T. Fleming (paperback, 6 × 9, 224 pages, $14.95)

Your Living Trust and Estate Plan: How to Maximize Your Family's Assets and Protect Your Loved Ones by Harvey J. Platt (paperback, 6 × 9, 304 pages, $14.95)